VIII International Conference for Young Scientists "High Technology: Research and Applications" (HTRA 2019)

IOP Conference Series: Materials Science and Engineering
Volume 754

Tomsk, Russia
30 September - 5 October 2019

ISBN: 978-1-7138-2660-6
ISSN: 1757-8981

Printed from e-media with permission by:

Curran Associates, Inc.
57 Morehouse Lane
Red Hook, NY 12571

Some format issues inherent in the e-media version may also appear in this print version.

Institute of Physics
Dirac House, Temple Back
Bristol BS1 6BE UK

Phone: 44 1 17 929 7481
Fax: 44 1 17 920 0979

techtracking@iop.org

Additional copies of this publication are available from:

Curran Associates, Inc.
57 Morehouse Lane
Red Hook, NY 12571 USA
Phone: 845-758-0400
Fax: 845-758-2633
Email: curran@proceedings.com
Web: www.proceedings.com

TABLE OF CONTENTS

IOP Conf. Series: Materials Science and Engineering **754** (2020) 011001 doi:10.1088/1757-899X/754/1/011001

Preface

The VIIIth International Conference for Young Scientists "High Technology: Research and Applications" (HTRA-2019) was held at Tomsk Polytechnic University, Tomsk, Russia on September 30 - October 5, 2019. The conference "HTRA" is held at TPU annually, starting from 2012, and brings together young scientists, graduate students and experienced scientists from Russia and other countries (Kazakhstan, Azerbaijan, Vietnam, China, Thailand, Syria).

HTRA-2019 is aimed at developing cooperation between Russian and foreign young scientists and students to conduct joint research of engineering science and high technology application. The conference focuses on modern technology and advances in science and technology based on an interdisciplinary approach to problem solving. The main topics of the conference were as follows:

1. Chemical technology and biotechnology;
2. Rational use of natural and technology-related mineral raw materials and water resources;
3. Photonic technologies and photo engineering;
4. Reliability of mechanical engineering and mechanic technologies;
5. Advanced materials and technologies in the space industry.

We appreciate contributions from all participants of HTRA-2019. We also want to thank sponsors for the financial support and Program Committee members for their efforts in organization and holding of the HTRA-2019 Conference.

Dr. Alexey N. Yakovlev,

Chair of HTRA-2019 Organizing Committee

Tomsk Polytechnic University

Peer review statement

All papers published in this volume of *IOP Conference Series: Materials Science and Engineering* have been peer reviewed through processes administered by the proceedings Editors. Reviews were conducted by expert referees to the professional and scientific standards expected of a proceedings journal published by IOP Publishing.

IOP Conf. Series: Materials Science and Engineering **754** (2020) 012001 doi:10.1088/1757-899X/754/1/012001

Experimental Study of the Cells Interaction on Linear and Nonlinear Operating Modes

P V Pisarev and K A Akhunzianova

Perm National Research Polytechnic University, Perm, Russia

E-mail: pisarev85@live.ru

Abstract. In this article, the results of SAS samples laboratory research on an interferometer with a normal sound wave incidence at sound pressure levels of 100 and 130 dB are presented. The mutual influence of prismatic resonators was revealed, which leads to a decrease in the resonators group acoustic efficiency.

1. Introduction

At present, the ecology of air transport has become the second most urgent problem highlighted by the International Civil Aviation Organization (ICAO), giving way only to flight safety. The dominant component of the aviation harmful effects on the environment is the noise created by aircraft. The most promising way to reduce the aircraft engine noise is to use resonant sound-absorbing structures (SAS) [1]. SASs are a combination of Helmholtz cells made in the form of a polymer composite honeycomb structure.

Prediction of the Helmholtz resonators acoustic characteristics and increasing the efficiency of broadband noise reduction is the subject of research conducted by many national and foreign authors. Of particular interest are the studies of resonator groups' interaction and their joint work. For example, in [2], the interaction of 2 or more cylindrical cells was studied. In [3], a numerical study of the closely spaced cylindrical cellsinteraction was carried out. The optimal distances between the cells, as well as the geometric characteristics of the neck, were identified. In studies [4, 5] A. Selamet and others found the influence of plane and non planar wave propagation on the resonant frequency and wave attenuation in Helmholtz spherical cells while changing cell sizes. The study of the closely located resonators acoustic efficiency was the subject of research by many authors. In [2, 6], an experimental study was made of the mutual influence of closely located cells. It was revealed that when the distance between the cells necks is less than 0.25 of the wavelength, the efficiency of the pair of resonators is less than that of a single one.

At the same time, in order to choose a more effective SAS design, there is a need to study the properties of a single cell and the cells joint work, including in non-linear modes.

This paper reports on the results of laboratory research of prismatic SAS samples on an interferometer with a normal sound wave incidence at sound pressure levels of 100 and 130 dB are presented.

2. Experimental determination of SAS samples acoustic characteristics

To conduct the research, a geometric model of samples containing prismatic cells (resonators) was developed. The geometric characteristics of the cells were taken as follows: resonator height 10 mm, hexagon side length 8 mm. The diameter of the neck and the height of the resonator neck are 2 and 1 mm, respectively. In addition, a sample was developed containing a pair of resonators close to each other. Geometric models are presented in Figure 1. Based on the developed geometric models, reference samples were made using additive manufacturing technology. The manufacturing process is described in detail in [7].

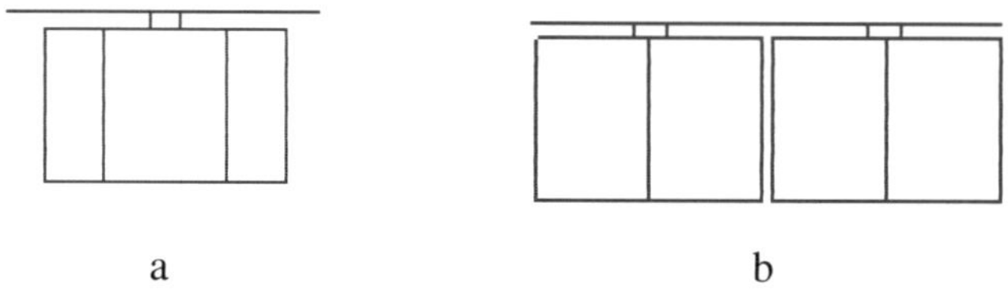

Figure 1. Samples geometric models: a - A1 model, b - A2 model

Acoustic tests in an interferometer with normal incidence of sound waves are widespread due to the relative simplicity of the experiment. The experimental setup is a tube of circular cross-section, on one end of which there is a SAS sample, on the other - a speaker that irradiates the sample with acoustic waves (Figure 2). Also, microphones which record the acoustic pressure of the incident and reflected waves in time are installed at a certain distance from the SAS sample in the channel of the interferometer flush with the wall (Figure 3). After that, the obtained pressure is processed by the fast Fourier transform algorithm, as a result of which the impedance of the SAS sample is calculated.

The impedance relates pressure and normal velocity at the boundary, the acoustic properties of which it characterizes. In most works, the influence of high sound pressure levels of the incident wave is described through the velocity of particles of the medium in the hole [8-12]. Most semiempirical models are based on the works of Crandall and Melling. In the Crandall's work [12], an expression was given for the impedance of a round hole, and in the Melling's work[13], the effect of high sound pressure levels, which leads to the SAS nonlinear behavior, was studied.

According to the definition, the dimensionless specific acoustic impedance is expressed as the ratio

$$Z = X + iY = \frac{p}{\rho c u_n}, \tag{1}$$

where p is the acoustic pressure; ρ is the air density; c is the speed of sound in air; u_n is the acoustic velocity.

Despite the apparent simplicity of expression (1), determining the impedance of a sound-absorbing structure in practice is a difficult task. Today, preference is given to the experimental determination of the impedance of one or another SAS sample.

Two microphones are used for the measurements, since the "transfer function method" used to determine the impedance is the easiest to calculate. At the same time, this option can be implemented if there is only a piston mode in the channel, which leads to a dependence of the installation frequency range on the channel dimensions of the impedance pipe.

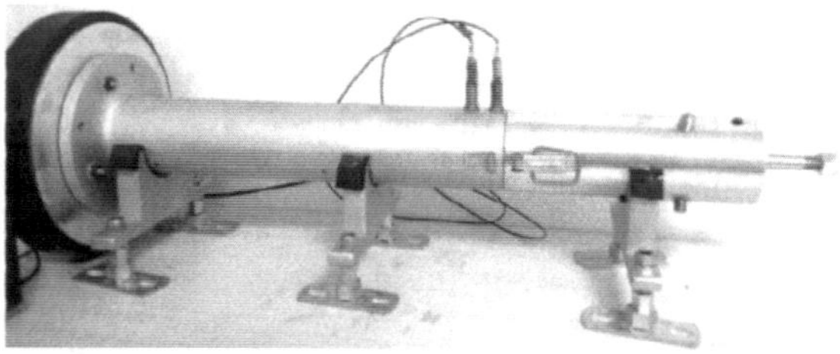

Interferometer Specifications:
- sound pressure up to 160 dB;
- frequency range 500-6400 Hz;
- size of test sample 30 mm.

Figure 2. Interferometer set and ready-to-use

IOP Conf. Series: Materials Science and Engineering **754** (2020) 012001 doi:10.1088/1757-899X/754/1/012001

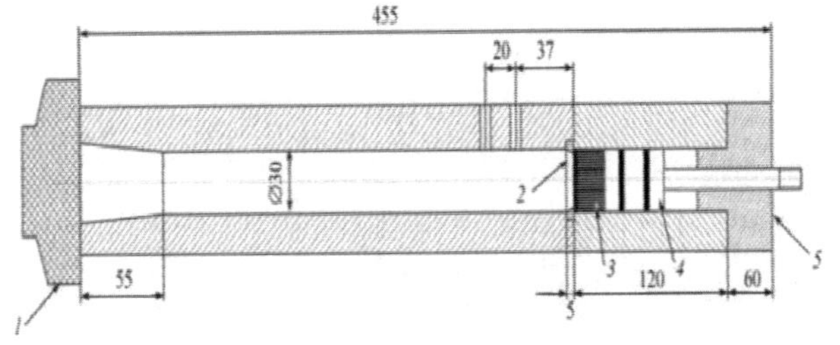

1 - speaker;
2 - washer fixing the sample;
3 –SAS sample;
4 - stock;
5 - directing plug.

Figure 3. The interferometer internal channel

In the course of experimental studies, the dependences of the sound absorption coefficient on frequency were obtained (Figure 4, 5). Figures 6, 7 show the dependences of the complex resistance (impedance) on frequency.

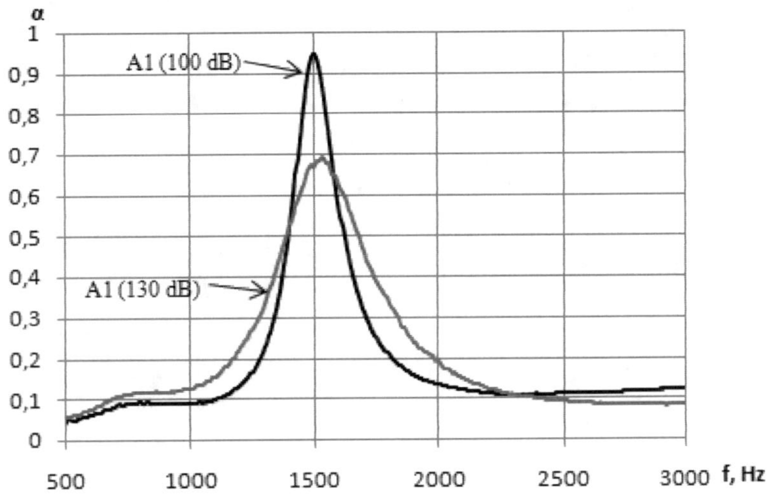

Figure 4. Graph of sound absorption coefficient versus frequency for a single resonator

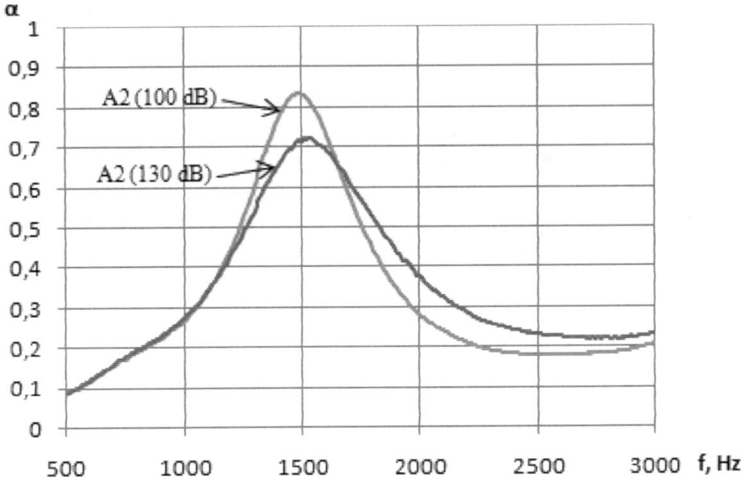

Figure 5. Graph of sound absorption coefficient versus frequency for a pair of resonators

IOP Conf. Series: Materials Science and Engineering **754** (2020) 012001 doi:10.1088/1757-899X/754/1/012001

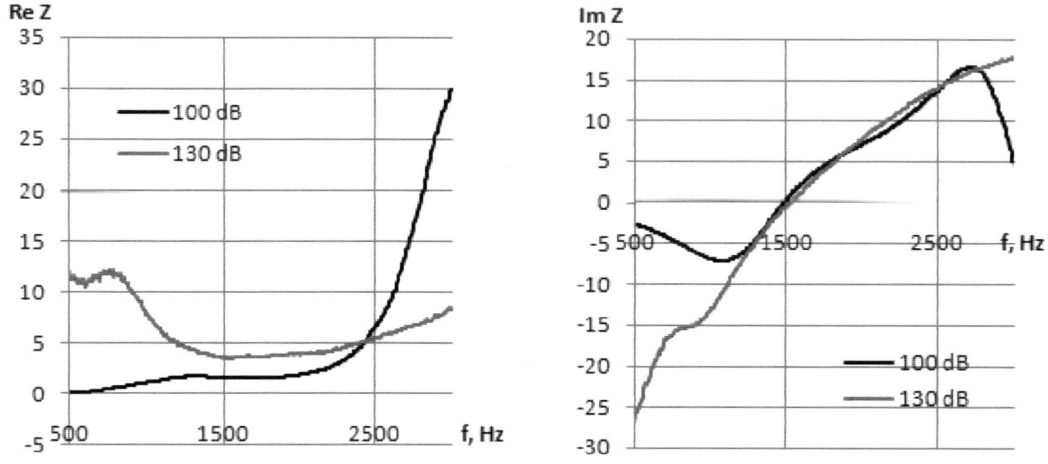

a b

Figure 6. The real (a) and imaginary (b) parts of the single resonator sample impedance

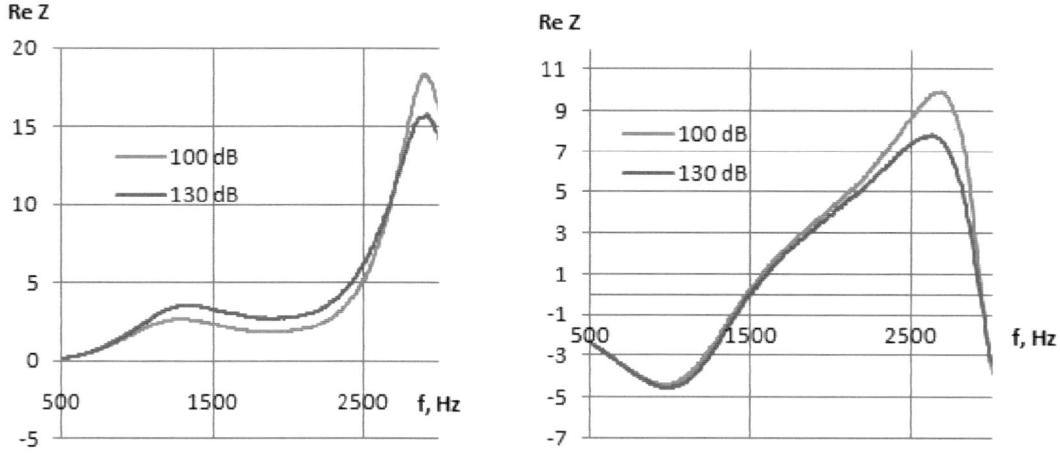

a b

Figure 7. The real (a) and imaginary (b) parts of the pair of resonators sample impedance

The analysis of the obtained dependences revealed that with an increase in the sound pressure level from 100 dB to 130 dB for A1 model, the sound absorption coefficient decreases by 26.8%.

The mutual influence of the cells during their joint operation was confirmed. When the model A2 resonators work together at a sound pressure level of 100 dB, the sound absorption coefficient decreases by 11.6%. And at a sound pressure level of 130 dB there is an increase in the sound absorption coefficient by 4.4% when compared with a single resonator. This indicates a decrease in the effect of the mutual influence of the cells in non-linear operating modes.

3. Conclusion

Experimental researches of the prismatic resonators acoustic efficiency during their joint work was carried out. It should be noted that, the mutual influence of prismatic resonators was revealed, which leads to a decrease in the group acoustic efficiency. When the A2 model resonators worked together at a sound pressure level of 100 dB, the sound absorption coefficient decreased by 11.6%. And at a sound pressure level of 130 dB there was an increase in the sound absorption coefficient by 4.4% when compared with a single resonator. The hypothesis of a decrease in the acoustic efficiency of a group of same volume prismatic cells during joint operation was confirmed.

Acknowledgments
The research was carried out at the Perm National Research Polytechnic University with the support of the Russian Science Foundation (project No. 18-79-00295).

References

[1] Zakharov A G, Anoshkin A N, Pan'kov A A and Pisarev P V 2016 PNRPU Aerospace Engineering Bulletin **46** 144-158
[2] Li D 2003 PhD Thesis University of Pittsburgh, USA
[3] Mahmud Md A, Hossainb Md Z, Islamc Md S and Morshedd Mir Md M 2015 Global Science and Technology Journal **3** 1 1–9
[4] Selamet A, Radavich P M, Dickey N S and Novak J M 1997 The Journal of the Acoustical Society of America **101** **1** 41-51
[5] Selamet A, Dicky N S and Novak J M 1995 Journal of sound and vibration **187** **2** 358-367
[6] Soh Y P, Yap E W T and Gan B H L 2001 Level 4 Thesis University of Adelaide, Australia
[7] Pisarev P V, Anoshkin A N and Merzliakova N A 2018 MATEC Web Conf. **243** 00026
[8] Sobolev A F 2007 Acoustic Journal **53** **6** 861–872
[9] Elnady T and Boden H 2003 AIAA Paper 2003–3304
[10] Yu J, Ruizand M and Kwan H W 2008 AIAA Paper 2008–2930
[11] Munin A G and Kvitka V E 1973 Aviation acoustics (Mechanical Engineering) p 448
[12] Crandall I 1972 Theory of vibrating systems and sound (D. Van Nostrand & Co) p 272
[13] Melling T 1973 Journal of Sound and Vibration **29** 1 1–65

IOP Conf. Series: Materials Science and Engineering **754** (2020) 012002 doi:10.1088/1757-899X/754/1/012002

Temperature Field Measurement in a Cutting Tool by Laser Interferometry

I A Efimovich and I S Zolotukhin

Institute of Industrial Technology and Engineering, Tyumen Industrial University, 38 Volodarskogo Street, Tyumen, 625000, Russian Federation

[1] E-mail: zolotuhinis@tyuiu.ru

Abstract. To increase machining efficiency, in addition to determining the temperature distribution on the tool surface, it is also necessary to determine the internal temperature distribution. The existing methods for determining the cutting tool temperature distribution and temperature field measurement have significant disadvantages, which limit their applicability and accuracy. The purpose of this study is to develop and test a novel cutting tool temperature field measurement method. The method involves recording the cutting tool thermal expansion fields by laser interferometry and calculating the tool material temperatures with its coefficient of thermal expansion (CTE). Compared with the methods employing infrared thermometry, the method developed in the present study has a higher spatial resolution and a lower achievable field of view due to its usage of light in the visible region. In addition, since the oxidation film has a higher reflection coefficient for visible light, the present method is less sensitive to the thin-film interference on an oxidized tool surface than those using infrared thermometry, eliminating the false shift of the measured temperature. Moreover, by utilizing cheaper optical components and equipment for interference fringe pattern registration, this method is more affordable. Unlike the emissivity coefficient in infrared thermometry, CTE is independent of changes in the surface quality and can be measured with high accuracy by modern dilatometry. Therefore, the developed method, which employs CTE to calculate the temperature, has a higher reliability. In the present study, the efficiency of this method is tested by the orthogonal turning of difficult-to-cut martensitic heat-treated steel with a cemented tungsten carbide tool. Through the experiment, the temperature distributions along the rake face and flank of the tool, as well as the temperature field inside the tool, were obtained. The results can improve the temperature field measurement in machining cutting tools.

1. Introduction

Temperature is one of the most important factors that influence various aspects of the metal cutting process. The thermal state of the cutting tool, which is one of the most important elements in machining, is extremely non-uniform. To increase machining efficiency, it is necessary to determine the temperature distribution on both the surface of the cutting tool as well as the interior. The study of temperature gradients allows the determination of thermal stresses and characteristics of thermal flows inside the tool. Moreover, it is useful for determining the tool material state and the nature of the contact processes in the cutting zone.

Currently, researchers focus on analytical methods of determining cutting tool temperature distribution. Idealized cutting zone models are used in those methods to calculate power and distribution character of the heat sources and sinks [1-3]. Moreover, the accuracy of temperature calculations varies widely, and a good result can be achieved for only a narrow range of cutting parameters. Due to the disadvantages of the analytical methods, only experimental methods can give the most reliable results at the present time.

The cutting zone has a small size and a relatively high power of heat sources. Therefore, the temperature field of the cutting tool can have a high gradient, and temperatures at some points can become very high. These properties cause difficulty in the experimental determination of temperature fields. There are, however, a number of experimental methods developed to attempt to measure the temperature distribution in the cutting tool.

The split tool method is a variation of the tool-work (natural) thermocouple method that provides an opportunity to determine temperature distributions along the rake and clearance faces, but not a temperature field. Moreover, since the tool consists of two parts composed of dissimilar materials with a gap between them, the properties of the split tool differ from the real cutting tool significantly.

A method employing embedded artificial and semi-artificial thermocouples [4, 5] that are distributed on the faces or into the body of tool differs in the complexity of the tool-making process, and it does not allow for temperature measurements near the cutting edge area.

The running thermocouple method [6] is a time-consuming method in terms of experiment preparation and cannot be used with conditions of high cutting speeds or low feed rates.

The method based on material microstructure and hardness changes [7] can be used for a tool composed of high-speed steel. The structure and hardness of the cemented carbide tool is not significantly affected by heating; therefore, this method is less applicable.

The thermal indicator methods [8], among which the physical vapor deposition (PVD) film and temperature-sensitive paint methods are notable, are of particular interest. However, high-gradient temperature fields obtained by these methods present low accuracy due to the thermophysical differences between the films and the tool material. Moreover, fields with a wide temperature range and short interval isotherms are difficult to study by those methods. The tempering colors method (oxide layer method) is not suitable for quantifying temperatures.

The infrared thermometry method currently gives the most accurate results and is therefore used most extensively. However, its accuracy highly depends on the measuring accuracy of the implemented emissivity coefficient, which can change with increasing temperature and depends on the roughness and oxidation state of surface. Due to thin film interference on oxide layers, the method can result in a false shift of the measured temperature. Moreover, infrared cameras are difficult to use for studying small objects due to their relatively low spatial resolution and small achievable field of view, caused by the long wavelength of infrared radiation [9, 10]. In addition, difficulties in measuring large temperature changes and the high cost of infrared detector arrays and optical elements determine the limited application of the method.

2. Method

To improve the process of measuring the cutting tool temperature fields and overcome the disadvantages of the above-mentioned methods, we developed a method [11] based on a laser interferometry.

This method is implemented for orthogonal cutting. When a cutting tool works, it heats and expands. Shifts and shape changes of the tool side surfaces are determined by changing interference fringe patterns obtained by a Fizeau-type interferometer, of which a simplified schematic is shown in figure 1. The interferometer is formed by surface A of the tool (1) and surface B of the wedge prism (2). The laser beam hitting the wedge prism is split into two new beams. The first (reference) beam is reflected from the surface B of the wedge. The second (measurement) beam is transmitted through the wedge and is reflected from the mirror side surface A of the tool. Both reflected beams recombine and produce an interference pattern, which represents the shape of the tool surface A. A feature of the applied interferometer design [12] is the rigid fastening of the optical wedge relative to the tool, which eliminates their relative shift induced by vibration during cutting and increases the interference pattern quality.

IOP Conf. Series: Materials Science and Engineering **754** (2020) 012002 doi:10.1088/1757-899X/754/1/012002

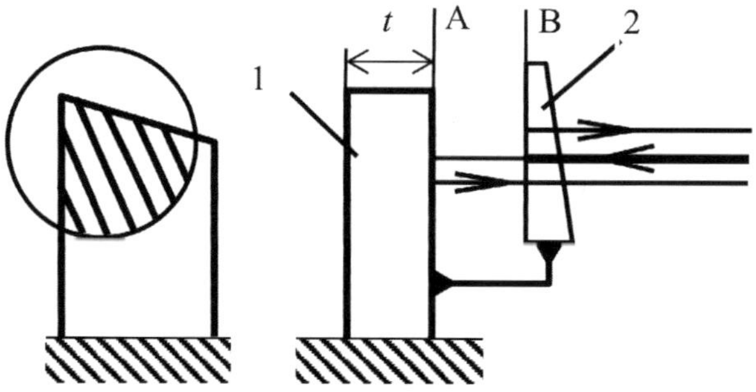

Figure 1. Schematic of interferometer.

Since the power of heat sinks on the tool side surfaces is low, and the tool width is narrow relative to its size, temperature distribution parallel to the cutting edge direction may be regarded as uniform. Therefore, due to symmetrical expansion of the tool side surfaces while heating, a tool width change Δt at the point of interest may be determined by:

$$\Delta t_i = \frac{\Delta m_i \cdot \lambda}{n},$$

where $\Delta m_i = m_t - m_0$ – difference of the interference fringe order at the point of interest between the heated and cold tools respectively; λ – laser wavelength; n – refractive index of the air.

Temperature change at the point of interest in the temperature field may be found by:

$$\Delta T_i = \frac{\Delta t_i}{t_0} \cdot \frac{1}{\alpha},$$

where t_0 – width of cold tool; α – mean coefficient of thermal expansion (CTE) for tool material, which can be measured by any dilatometry method or found in reference literature [13].

Temperature at the point of interest may be found by:

$$T_i = \Delta T_i + T_0,$$

where T_0 – temperature of a cold tool.

To obtain the temperature distribution along the rake and clearance faces, fringe points along the corresponding lines have to be analyzed. Temperature fields may be determined using either of two methods: the first method involves the analysis of every fringe point inside the region of interest (ROI). However, since there are no heat sources on the tool side surfaces in the ROI and heat sinks have a low power, the tool thermal state during the steady cutting process or for a relatively short period of time may be depicted by the two dimensional steady-state case. The second method involves the analysis of fringe pattern points on the closed boundary of the ROI (e.g. bounded by lines) and calculating the temperature at points inside the region by solving the Laplace equation:

$$\nabla^2 T = 0 \qquad (1)$$

The equation may be solved by any computer-based numerical method. The second method is preferable due to its low sensitivity to the fringe pattern quality.

3. Experimental Rig

Figure 2 shows a schematic of the experimental rig used to test our developed method. A beam from a laser (1) passed through a beam expander (2) and a cubic beam splitter (3), strikes an interferometer, which comprised an optical wedge (4) and a cutting tool (5) with a polished side surface (6), rigidly fastened relative to one another by a tool holder (7). After interference of the reference and measurement beams, the resulting beam reflected on the beam splitter's diagonal surface was directed to a high-speed camera (8) lens. The interference fringe patterns recorded by the camera were saved on a computer (9).

The experimental rig was mounted on a conventional lathe with an adjustable speed spindle drive (figure 3). On the cross slide of the lathe, the base table with the tool holder (1) was placed with two

IOP Conf. Series: Materials Science and Engineering **754** (2020) 012002 doi:10.1088/1757-899X/754/1/012002

aluminum rails fixed on it. A single-frequency DPSS laser (2)(LCM-S-111 (λ=532 nm)), the beam expander (3), and the rotation stage with a cubic beam splitter (4) were mounted on one rail. The second rail carried the high-speed camera (5)(Fastec HiSpec 2-HR with zoom lens (6) NAVITAR Zoom 6000). Every optical element had an antireflective coating.

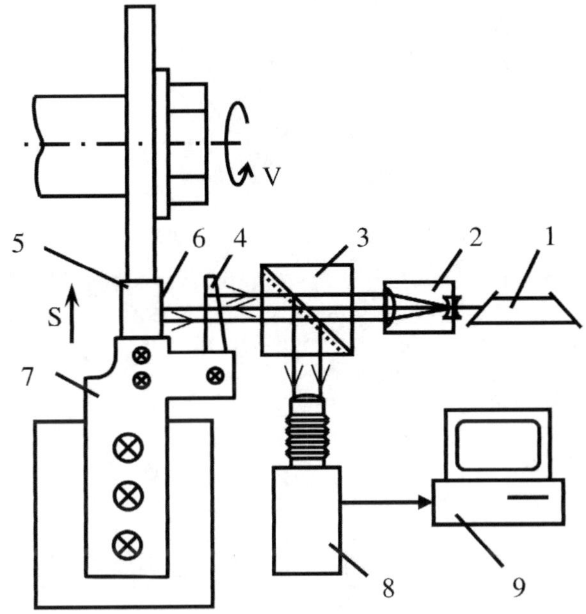

Figure 2. Schematic of experimental rig.

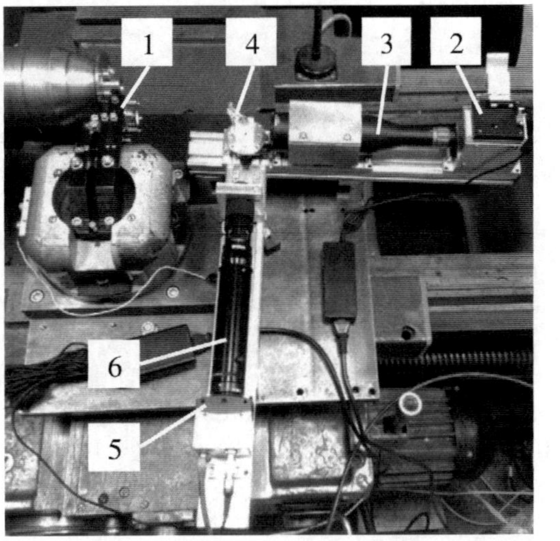

Figure 3. View of experimental rig.

4. Experiment conditions

To confirm the efficiency of this method, we tested it with the following machining parameters: the workpiece was a 4.2-mm thick disc composed of difficult-to-cut martensitic heat-treated steel grade X13Cr11Ni2W2MoV [14], the chemical composition of which is presented in table 1. This steel is used for the production of turbine disks and blades. The steel hardness was HB 277.

Table 1. Chemical composition of steel grade X13Cr11Ni2W2MoV.

Element	C	Cr	Ni	W	Mo	V	Si	Mn	S	P	Fe
wt %	0.10 - 0.16	10.5 - 12.0	1.5 - 1.8	1.6 - 2.0	0.35 - 0.50	0.18 - 0.30	max 0.6	max 0.6	max 0.025	max 0.03	bal.

The cutting tool was composed of grade VK8 [15] cemented tungsten carbide, with 8% cobalt content by weight, and had a 0° rake angle, 8° clearance angle, and 4.6-mm width. Cutting speed V and feed rate S were 6 m/min and 0.14 mm/rev respectively.

5. Results and Discussion

Figure 4 shows the interference fringe patterns for a cold tool (a) with T_0=20°C, and a heated tool (b) right after the interruption of the cutting process.

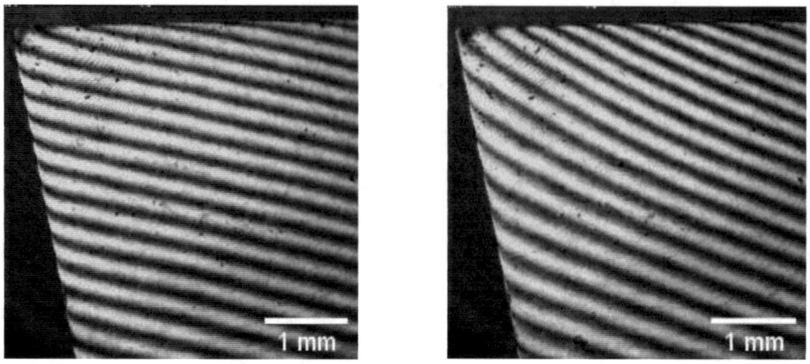

a b

Figure 4. Interference fringe patterns for cold (a) and heated (b) tool.

To obtain the fringe order distribution plot, we generated interference pattern profiles along the rake and the clearance faces in ImageJ, and then found maxima of image intensity for the cold and heated tools. Every maximum was numbered relative to a randomly selected fringe, which had a position on both patterns that could be recognized accurately. The most convenient position for this purpose was the farthest from the heating zone fringe. The plot shows the number of profile maxima (fringe order) and their positions relative to the cutting edge. To obtain values for tool expansion caused by heating, the plots for the warm and cold tools have to be subtracted. Figure 5 shows fringe order distributions along the rake (a) and clearance (b) faces for heated and cold tools and their difference.

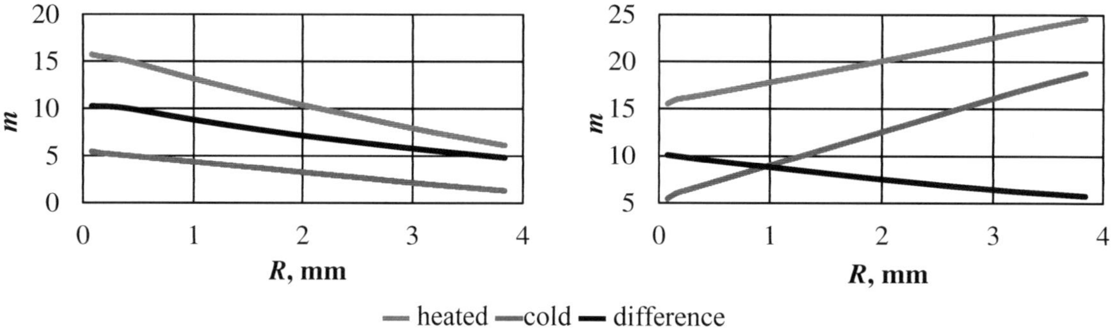

a b

Figure 5. Fringe order m distributions along the rake (a) and clearance (b) faces with respect to the distance R from the cutting edge.

To convert fringe order distribution to temperature distribution, the CTE of the tool material [16] was used. Figure 6 shows temperature distributions along the rake (a) and clearance (b) faces. Plotting and mathematical processing was performed using MS Excel. Figure 7 shows the temperature field, which was obtained by solving the Laplace equation (1) in Cartesian coordinates via the finite difference method, computed by a specially-created program in Turbo Basic. C and C_l on the figure show the length of the tool in contact with the chip and the workpiece respectively.

6. Conclusion

We developed an experimental method for measuring temperature fields, which involves recording the cutting tool thermal expansion fields by laser interferometry and calculating the tool material temperatures with its CTE.

Compared to the methods employing infrared thermometry, the method developed in this study demonstrates a higher spatial resolution and a lower achievable field of view due to its use of light in the visible region. Moreover, by utilizing cheaper optical components and equipment for interference fringe pattern registration, this method is more economical.

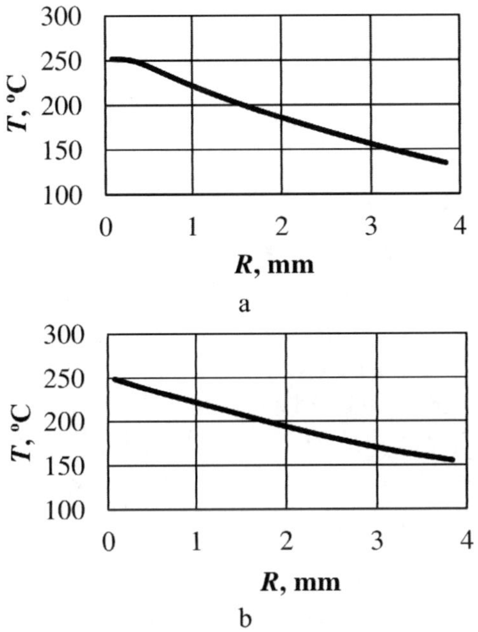

Figure 6. Temperature *T* distribution along the rake (a) and clearance (b) faces with respect to the distance *R* from the cutting edge.

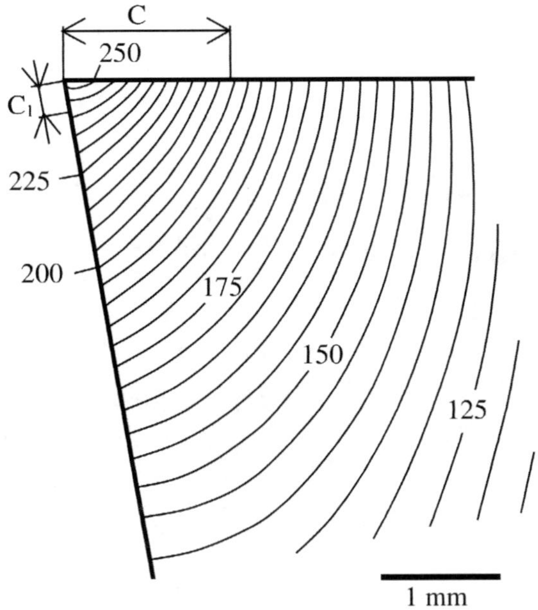

Figure 7. Temperature field of heated tool (5 °C isotherm interval).

Furthermore, since the oxidation film has a higher reflection coefficient for visible light, the proposed method is less sensitive to the thin-film interference on an oxidized tool surface than methods that employ infrared thermometry; this eliminates the false shift of the measured temperature.

Unlike the emissivity coefficient in infrared thermometry, the CTE is unaffected by changes in the surface quality and can be measured with high accuracy by modern dilatometry. Therefore, the developed method, which employs CTE to calculate the temperature, is more reliable.

References

[1] Komanduri R and Hou Z B 2001 *Int. J. Mech. Sci.* **43** 89–107
[2] Abukhshim N A, Mativenga P T and Sheikh M A 2005 *Int. J. Mach. Tool. Manu.* **46** 782-800
[3] Klocke F, Brockmann M, Gierlings S and Veselovac D 2015 *Mech. Sci.* **6** 89-94
[4] Komanduri R and Hou Z B 2001 *Tribol. Int.* **34** 653–82
[5] Davies M A, Ueda T, M'Saoubi R, Mullany B and Cooke A L 2007 *CIRP Ann.-Manuf. Techn.* **56** 581-604
[6] Astakhov V P 1998 *Metal Cutting Mechanics* (Boca Raton: CRC Press) p 320
[7] Trent E M and Paul K W 2000 *Metal cutting* 4th ed. (Boston: Butterworth-Heinemann) p 464
[8] Longbottom J M and Lanham J D 2005 *Aircr. Eng. Aerosp. Tec.* **77** 122-30
[9] Pujana J, del Campo L, Pérez-Sáez R B, Tello M J, Gallego I and Arrazola P J 2007 *Meas. Sci. Technol.* **18** 3409
[10] Daniels A 2018 *Field Guide to Infrared Systems, Detectors, and FPAs* 3rd ed. (Bellingham: SPIE Press) p 170
[11] Efimovich I A, Zolotukhin I S and Shvetsova E I 2010 *RF Patent Specification* 2442967
[12] Efimovich I A 2010 *RF Patent Specification* 2423663
[13] James J D, Spittle J A, Brown S G R and Evans R W 2001 Meas. Sci. Technol. **12** R1–R15
[14] Russian Federation State Standard 1997 GOST 5632–72
[15] Russian Federation State Standard 1998 GOST 3882–74
[16] Efimovich I A, Zolotukhin I S and Zav'yalov E S 2019 *Obrabotka metallov* **21** 129–40

IOP Conf. Series: Materials Science and Engineering **754** (2020) 012003 doi:10.1088/1757-899X/754/1/012003

Boundary element analysis of dynamic interaction of anisotropic elastic half-space and structure under impact load

I P Markov and L A Igumnov

Research Institute for Mechanics, National Research Lobachevsky State University of Nizhny Novgorod, 23, bldg. 6, Prospekt Gagarina (Gagarin Avenue), Nizhny Novgorod, 603950, Russia

E-mail: teanku@gmail.com

Abstract. In this paper, a direct boundary element formulation for numerical modelling of three-dimensional dynamic soil-structure interaction is briefly described. Formulation is applicable to problems involving isotropic and anisotropic elastic materials. It is based on the Laplace domain regularized boundary integral equations for the displacements under assumption of vanishing initial conditions and zero body forces. Mixed boundary elements are employed for the spatial discretization of the boundary integral equations. Subdomain technique is used to tackle problems involving piecewise homogeneous material properties. The described formulation is employed for the boundary element analysis of dynamic soil-structure interaction problem involving a structure which contains a cavity with a complex geometrical form. Soil is represented by a transversely isotropic elastic half-space. Effect of material anisotropy of half-space on the transient response of the embedded structure is assessed. It is observed that the shear modulus anisotropy have substantial impact on the displacements response. Influence of the Young's modulus in the plane of isotropy is lesser but still pronounced.

1. Introduction

Consideration of the dynamic interaction between the supporting soil and the embedded structure is essential for ensuring safety, reliability and integrity of complex structures, buildings and foundations subjected to dynamic loadings like blast or impact. Naturally occurring soils often exhibit anisotropic behavior. Representing soil in practical problems as an isotropic medium can lead to inadequate or unsatisfactory results. Boundary Element Method (BEM) is well-known as an efficient and accurate numerical method for modelling transient wave propagation is linearly elastic solids and media, but the number of works on the BEM applied to dynamic problems which involves anisotropic elastic soils is quite low [1-8]. Primarily it is attributed to the difficulties associated with numerical implementation of the dynamic fundamental solutions of anisotropic elasticity, since the numerical integration of double integrals has to be performed for each pair of field and source points.

In this work we briefly describe a direct boundary element formulation for numerical analysis of three-dimensional dynamic soil-structure interaction problems and apply it to a problem which involves a protective cover structure of a nuclear heating plant (nuclear plant which produce thermal energy for district heating) partially embedded into a transversely isotropic soil. Effect of anisotropic characteristics of the soil on transient response of the structure is studied.

2. Boundary element formulation

The Laplace domain equations of motion for a homogeneous linearly elastic solid $\Omega \in R^3$ bounded by a surface $\Gamma = \partial\Omega$ assuming zero initial conditions and absence of the body forces are

$$\bar{\sigma}_{ij,j}(\mathbf{x},s) - \rho s^2 \bar{u}_i(\mathbf{x},s) = 0, \ \ \mathbf{x} \in \Omega, \ \ i,j = \overline{1,3}, \tag{1}$$

where $\bar{\sigma}_{ij}$ is the stress tensor, ρ is the mass density, s is the Laplace transform parameter and $\bar{u}_i$ are the displacements.

The symmetric linear strain tensor $\bar{\varepsilon}_{ij}$ is defined from the strain-displacement relationship

$$\bar{\varepsilon}_{ij}(\mathbf{x},s) = \frac{1}{2}\left(\bar{u}_{i,j}(\mathbf{x},s) + \bar{u}_{j,i}(\mathbf{x},s)\right). \tag{2}$$

The stress tensor and the strain tensor are linearly related through the generalized Hooke's law via the fourth-order elasticity tensor C_{ijkl} as follows

$$\bar{\sigma}_{ij}(\mathbf{x},s) = C_{ijkl}\bar{\varepsilon}_{kl}(\mathbf{x},s), \ \ k,l = \overline{1,3}. \tag{3}$$

Combining equations (1)-(3) and symmetry of stress tensor the linear field equations are obtained

$$C_{ijkl}\bar{u}_{k,lj}(\mathbf{x},s) - \rho s^2 \bar{u}_i(\mathbf{x},s) = 0, \ \ \mathbf{x} \in \Omega. \tag{4}$$

Problem is supplemented with an appropriate boundary conditions

$$\bar{u}_i(\mathbf{x},s) = \bar{u}_i^*(\mathbf{x},s), \ \ \mathbf{x} \in \Gamma_u, \tag{5}$$

$$\bar{t}_i(\mathbf{x},s) = \bar{\sigma}_{ik}(\mathbf{x},s)n_k(\mathbf{x}) = \bar{t}_i^*(\mathbf{x},s), \ \ \mathbf{x} \in \Gamma_t, \tag{6}$$

where n_k denote components of the outward unit normal vector to the boundary Γ at point $\mathbf{x}$.

The Laplace domain regularized direct boundary integral equations (BIEs) for displacements corresponding to the differential equations (4) can be expressed as follows

$$\int_{\Gamma}\left[\bar{u}_k(\mathbf{y},s)\bar{h}_{jk}(\mathbf{y},\mathbf{x},s) - \bar{u}_k(\mathbf{x},s)h_{jk}^S(\mathbf{y},\mathbf{x}) - \bar{t}_k(\mathbf{y},s)\bar{g}_{jk}(\mathbf{y},\mathbf{x},s)\right]d\Gamma(\mathbf{y}) = 0, \ \ \mathbf{x} \in \Gamma, \tag{7}$$

where $\mathbf{x}$ is the field point and $\mathbf{y}$ is source point, $\bar{g}_{jk}$ and $\bar{h}_{jk}$ are three-dimensional Laplace domain displacement and traction dynamic full-space fundamental solutions, respectively, h_{jk}^S is the singular term of $\bar{h}_{jk}$.

In the present formulation, so-called mixed boundary elements [9] are employed for the spatial discretization of BIEs (7). The geometry of a mixed quadrangular boundary element is interpolated using quadratic shape functions. Linear variation of displacements and a constant value of tractions are assumed over each boundary element. For the problems with anisotropic materials, fundamental solutions are expressed in terms of finite double integrals [10, 11]. Special attention is paid to the numerical evaluation of these integrals due to the highly oscillatory nature of the integrand for large values of the imaginary part of the complex frequency s.

Applying nodal collocation procedure, carrying out necessary integrations and taking into account prescribed boundary conditions we obtain a complex-valued system of linear algebraic equations for a given value of the Laplace transform parameter s. For the problems involving piecewise homogeneous material properties (e.g. soil-structure interaction) we employ a subdomain technique: boundary element procedure is repeated for each subdomain in which the material properties are constant and the coupling constraints follow from the conditions of equilibrium of tractions and continuity of the displacements at the interfaces between the subdomains.

3. Numerical analysis

The described formulation is employed to conduct the boundary element analysis of complex soil-structure interaction problem involving a protective cover structure of a nuclear heating plant partially embedded into a soil which is represented by a homogeneous anisotropic elastic half-space. Symmetrical part (for $x_2 \geq 0$ m) of the considered problem's geometrical configuration is depicted in figure 1. Protective cover structure is represented by an isotropic elastic rectangular parallelepiped with dimensions $L \times L \times H$ and it contains a cavity with a complex geometrical form as shown in detail in figure 2. The values of constants indicated in figure 2 are given as follows: $H = 34.5$ m, $L = 15.6$ m, $H_e = 14.455$ m, $l_1 = 9.6$ m, $l_2 = 11.6$ m, $h_1 = 2$ m, $h_2 = 9.3$ m, $h_3 = 4.215$ m, $h_4 = 4.01$ m, $h_5 = 8.445$ m, $h_6 = 2$ m, $h_7 = 2.2$ m, $h_8 = 0.33$ m, $r_1 = 3.6$ m, $r_2 = 3.8$ m, $r_3 = 4.29$ m, $r_4 = 5.8$ m and point $O(0,0,0)$ m is the origin of a rectangular Cartesian coordinate system $\{O; x_1, x_2, x_3\}$. On the area specified by relations $0\text{ m} \leq x_1 \leq 7.8$ m, $0\text{ m} \leq x_2 \leq 7.8$ m, $x_3 = 20.045$ m, the cover structure is subjected to an impact loading $t_3(t) = -t_3^* H(t)$, $t_3^* = 1 \cdot 10^7$ Pa, $H(t)$ is the Heaviside step function. The half-space and the rest of the protective cover structure are considered to be free of tractions. The elastic properties of the protective cover structure are as follows: $E_{struct} = 3 \cdot 10^{10}$ Pa, $\nu_{struct} = 0.2$, $\rho_{struct} = 2000\ \text{kg/m}^3$.

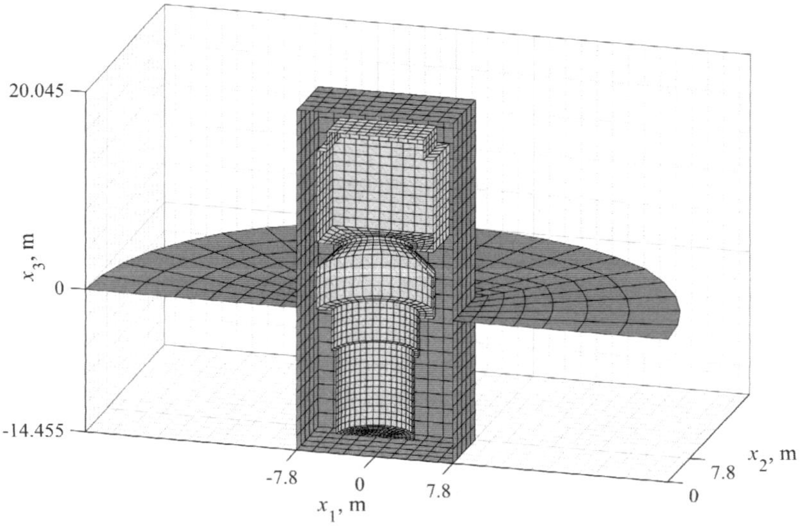

Figure 1. Configuration of the soil-structure interaction problem.

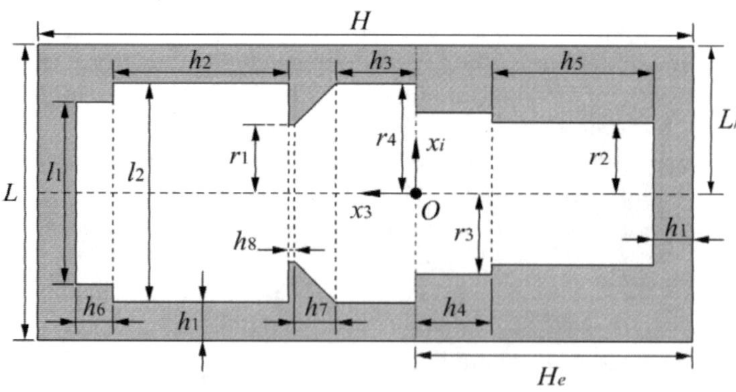

Figure 2. Geometry of protective cover structure.

To assess the effect of material anisotropy of the half-space on the transient response of the embedded structure we employ one artificial isotropic and several transversely isotropic materials. Density of all these fictitious materials is $\rho = 2000\ \text{kg/m}^3$. Young's modulus of isotropic material is $E_{iso} = 2\cdot 10^{10}$ Pa and Poisson's ratio is $\nu_{iso} = 0.25$. Planes of isotropy of transversely isotropic materials are parallel to $x_1 - x_2$ plane and x_3-axis is assumed to be the axis of rotational material symmetry. Non-zero elastic moduli in Voigt notation c_{ij} for a transversely isotropic soil can be expressed through the five independent engineering elastic constants E, E', ν, ν' and G' by

$$c_{11} = c_{22} = \frac{E\left(1-\eta\nu'^2\right)}{(1+\nu)\left(1-\nu-2\eta\nu'^2\right)},\quad c_{12} = c_{21} = \frac{E\left(\nu+\eta\nu'^2\right)}{(1+\nu)\left(1-\nu-2\eta\nu'^2\right)},\quad c_{44} = c_{55} = G', \quad (8)$$

$$c_{13} = c_{23} = c_{31} = c_{32} = \frac{E\nu'}{1-\nu-2\eta\nu'^2},\quad c_{33} = \frac{E'(1-\nu)}{1-\nu-2\eta\nu'^2},\quad c_{66} = \frac{E}{2(1+\nu)},\quad \eta = \frac{E}{E'}, \quad (9)$$

where E and ν are the Young's modulus and the Poisson's ratio in the $x_1 - x_2$ symmetry plane; E' and G' is the Young's modulus and the shear modulus in the x_3 – direction; ν' is the Poisson's ratio characterizing the lateral strain response in the $x_1 - x_2$ plane to a stress applied in the x_3 – direction.

For the problem under consideration we chose $E' = 2\cdot 10^{10}$ Pa, $\nu = \nu' = 0.25$. Properties of all materials involved in analysis are summarized in Table 1.

Table 1. Properties of materials for half-space.

			Elastic moduli, GPa					
Material	$\frac{E}{E'}$	$\frac{G'}{E'}$	c_{11}	c_{12}	c_{13}	c_{33}	c_{44}	c_{66}
Material 1	1	0.1	24	8	8	24	2	8
Material 2	1	0.2	24	8	8	24	4	8
Material 3	1	0.3	24	8	8	24	6	8
Material 4 (isotropic)	1	0.4	24	8	8	24	8	8
Material 5	2	0.4	56	24	20	30	8	16
Material 6	3	0.4	104	56	40	40	8	24
Material 7	5	0.4	440	360	200	120	8	40

Figures 3 and 4, 5 and 6 demonstrate transient responses of displacements $u_2(t)$ and $u_3(t)$ at the points $A(4.8,0,18.045)$m and $B(3.8,0,\text{-}4.01)$m located on the surface of the cavity inside the protective cover structure, respectively. Even though point A is located far above the surface of the half-space and very near the loading area, material parameters of the transversely isotropic half-space have noticeable influence on the displacements response. Much more pronounced impact on the response of the structure is evident from figures 5 and 6. In all cases, increasing ratio G'/E' and increasing ratio E/E' differently affects absolute values of the maximal displacements. Comparing to the solution for the isotropic half-space, shear modulus anisotropy have substantial influence on the vertical displacements $u_3(t)$ and lesser on the displacements $u_2(t)$. In contrast, the influence of the Young's modulus in the plane of isotropy is more notable for $u_2(t)$ than for $u_3(t)$, but altogether less than effect of E/E'. In addition, for a static formulation of the problem, figure 7 displays changes in distribution of displacements u_1 over a part of the cavity located in an embedded section of a structure. Obtained results convincingly illustrate the necessity to take into account material anisotropy of the embedding medium when solving static and dynamic soil-structure interaction problems.

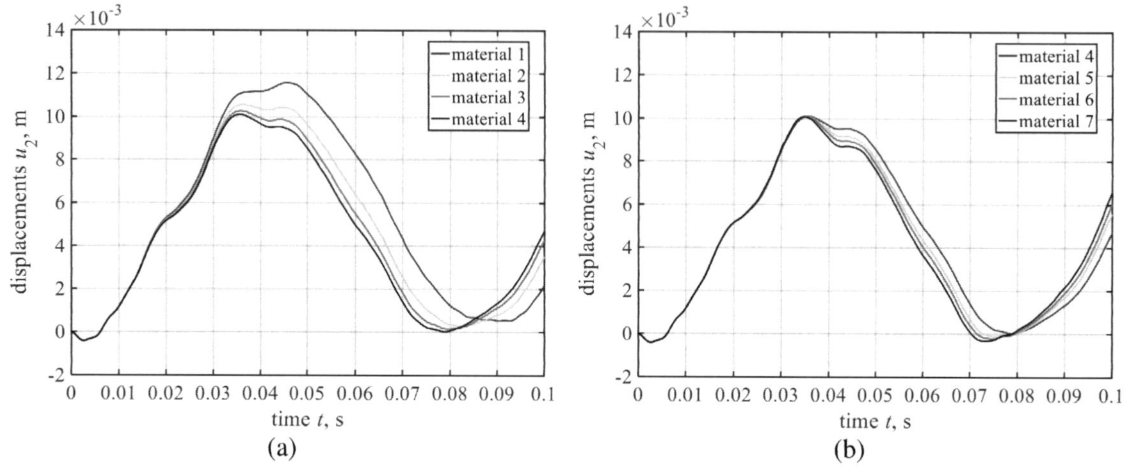

Figure 3. Displacements $u_2(t)$ at point *A* for (a) materials 1-4, (b) materials 4-7.

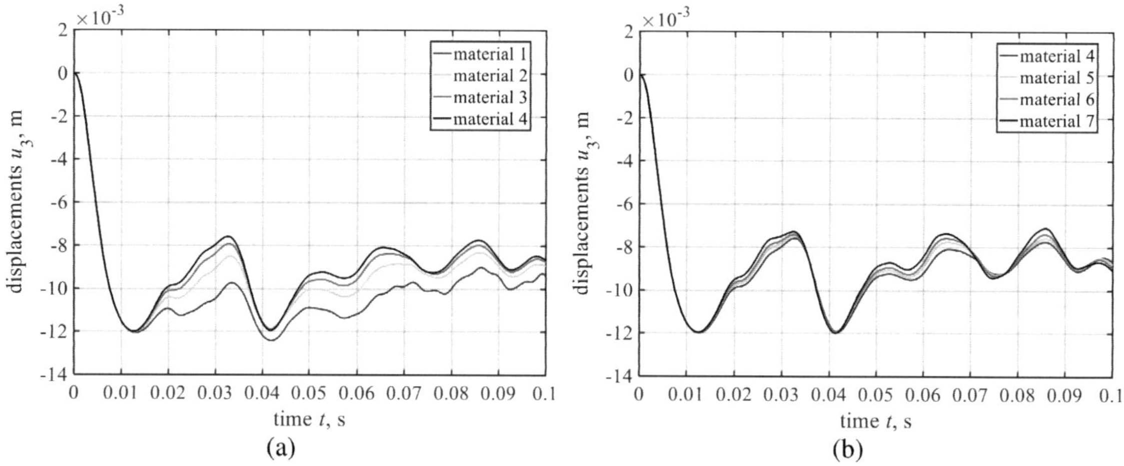

Figure 4. Displacements $u_3(t)$ at point *A* for (a) materials 1-4, (b) materials 4-7.

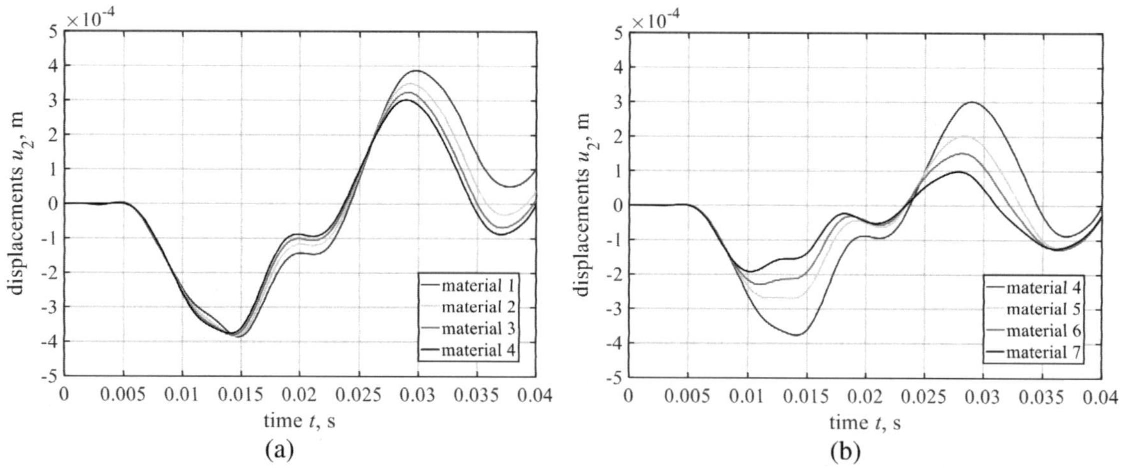

Figure 5. Displacements $u_2(t)$ at point *B* for (a) materials 1-4, (b) materials 4-7.

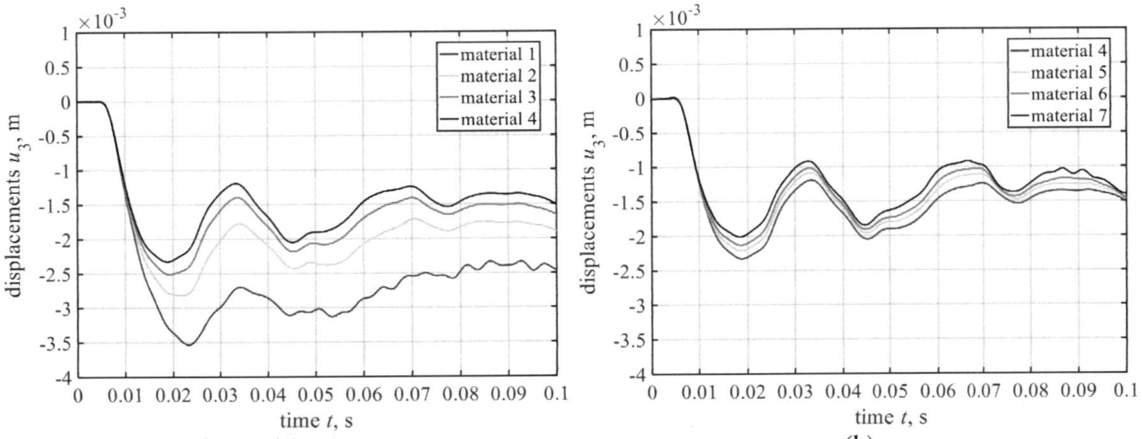

(a) (b)

Figure 6. Displacements $u_3(t)$ at point B for (a) materials 1-4, (b) materials 4-7.

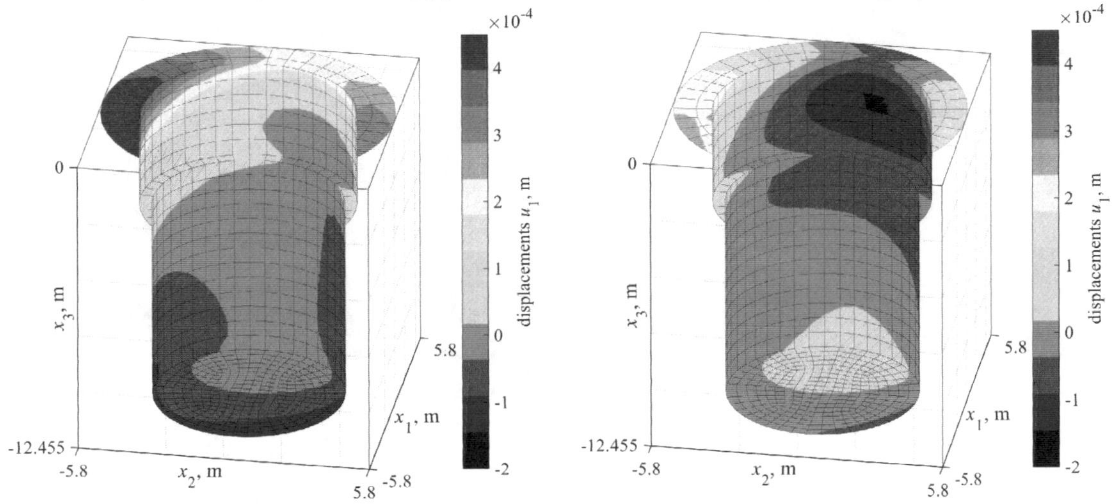

(a) (b)

Figure 7. Distribution of displacements u_1 for (a) material 1, (b) material 4.

Acknowledgments
The work is financially supported by the Russian Science Foundation under grant No. 18-79-00082.

References

[1] Pak R Y S and Guzina B B 1999 *Int. J. Solids Struct.* **36** 4743-4766
[2] Ahmad S, Leyte F and Rajapakse R K N D 2001 *J. Eng. Mech.* **27** 149-156
[3] Niu Y and Dravinski M 2003 *Wave Motion* **38** 165-175
[4] Niu Y and Dravinski M 2003 *Int. J. Numer. Methods Eng.* **58** 979-998
[5] Chuhan Z, Yuntao R, Pekau O A and Feng J 2004 *J. Eng. Mech.* **130** 105-116
[6] Barros P L A 2006 *Int. J. Numer. Anal. Methods Geomech.* **30** 683-702
[7] Furukawa A, Saitoh T and Hirose S 2014 *Eng. Anal. Bound. Elem.* **39** 64-74
[8] Ba Z and Gao X 2017 *Shock. Vib.* art. no. 2834274
[9] Igumnov L A, Karelin I S and Petrov A N 2011 *Probl. Strength Plast.* **73** 97-103
[10] Wang C Y and Achenbach J D 1994 *Geophys. J. Int.* **118** 384-392
[11] Wang C Y and Achenbach J D 1995 *Proc. R. Soc.* A **449** 441-458

Method of luminescent analysis of the state of photosynthetic apparatus of plants

D V Bylkov, S B Turanov and V I Korepanov

Division for Material Science, School of Advanced Manufacturing Technologies, National Research Tomsk Polytechnic University, 30 Lenin Avenue, Tomsk 634050, Russia

Email: prototype103faza@gmail.com

Abstract. Photosynthetic apparatus and its reaction to changes in external conditions carry information about the state of the plant. Thus, under different lighting conditions, as well as with changes in nutrition and microclimate conditions, the amount and composition of chlorophyll in plant cells changes, which will lead to changes in the optical parameters of plant chlorophyll. You can track the state of the plant by changing the parameters of chlorophyll fluorescence. To monitor the state of the photosynthetic apparatus, the method of analysis of induction curves of variable fluorescence is used. The purpose of this article is to design a research facility for testing the technique of fluorescent analysis and further accumulation of statistical data on the dependence of fluorescence characteristics on lighting parameters.

1. Introduction

At the moment, greenhouse complexes and its lighting systems are developing very actively. In the process of photosynthesis, which is used by plants to convert electromagnetic radiation (light), plants produce chemical energy necessary for their growth. The decisive factor for successful cultivation is lighting, using of adaptive LED irradiation systems is the efficiency increasing solution. Knowing what the plant needs throughout the growing season, you can develop an adaptive lighting system that will change the luminous flux and spectral composition, selecting the necessary parameters depending on the phase of growth. You can see it by using the technique of analyzing the state of the photosynthetic apparatus in the plant. By analyzing the fluorescence intensity of the plant, it is possible to monitor the reaction of the photosynthetic apparatus to certain changes in the irradiation modes, which will allow you to choose the necessary parameters of the irradiation mode.

During photosynthesis, not all of the light energy consumed by the plant is converted into chemical energy. Part of the energy is lost in the form of heat and the other in the form of fluorescence. Fluorescence is emitted mostly by chlorophyll *a*, it is a secondary emission of light energy absorbed by the chlorophyll molecule [1, 2].

IOP Conf. Series: Materials Science and Engineering **754** (2020) 012004 doi:10.1088/1757-899X/754/1/012004

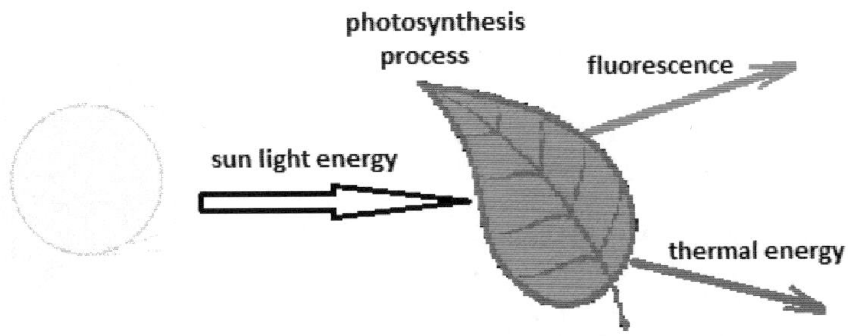

Figure 1. Process of energy redistribution in chlorophyll

Chlorophyll fluorescence *a* is a measure of photons unused during photosynthesis. In conditions of normal and effective work fluorescence remains on enough low level, and any violation of photosynthesis and lowering its efficiency to fast increasing fluorescence level. [1, 2]. Thus, by analyzing the quantum yield of fluorescence it is possible to judge the effectiveness of photosynthesis of the plant.

Plants adapted to the dark, irradiated by photosynthetically active radiation, emit fluorescence with a time-varying intensity, i.e. with a certain dynamic, called the induction curve of variable fluorescence [1].

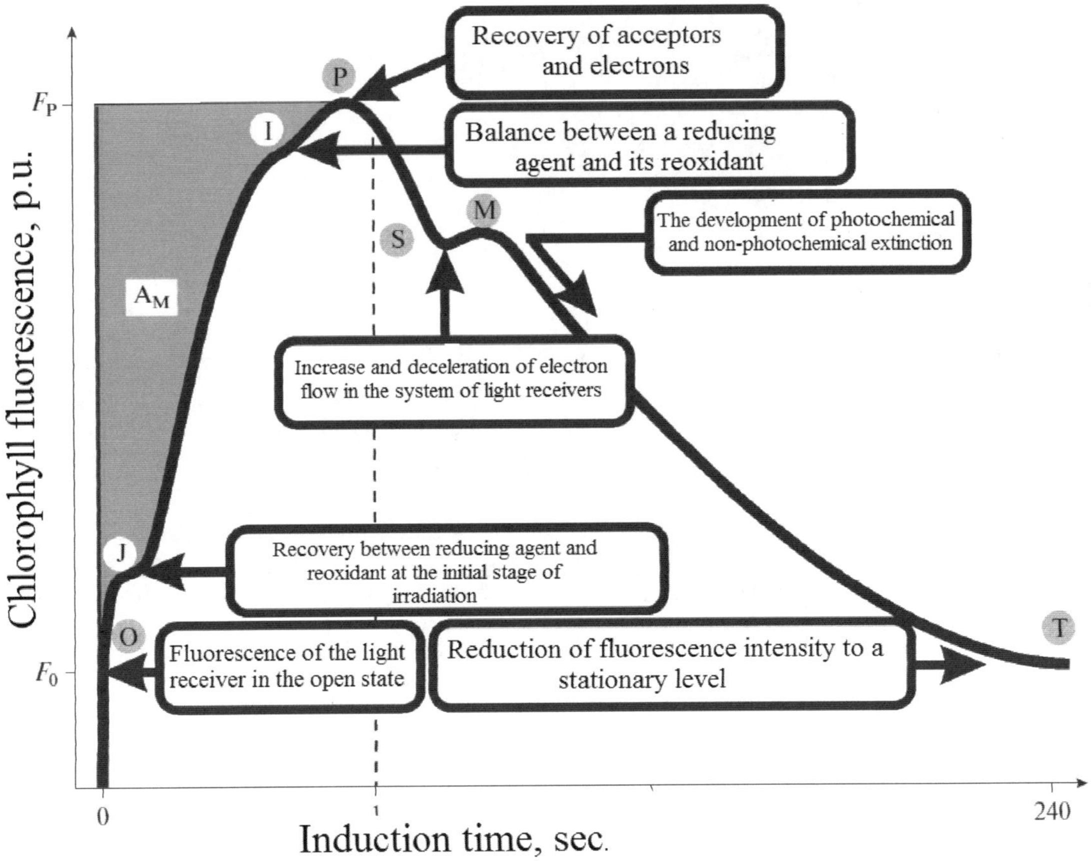

Figure 2. Typical induction fluorescence curve with an explanation of the processes of change in the fluorescence degree [1].

This curve describes the processes, which lead to changes in the level of fluorescence under different irradiation modes [1]. This technique is called JIP test (JIP – reference points of the induction curve) [1].

2. Materials and methods

This paper presents two methods for measuring chlorophyll fluorescence, namely a direct method for measuring the level of fluorescence and using amplitude modulation.

In the direct method, the pre-adapted to the dark plant is illuminated for 20-30 minutes by a constant LED light source, which has the function of changing the intensity and spectral composition of irradiation. The sample is illuminated by continuous light with an excitation wavelength in the range from 420 to 670 nm, the spectrofluorometer registers fluorescence light at wavelengths equal to 680-760 nm.

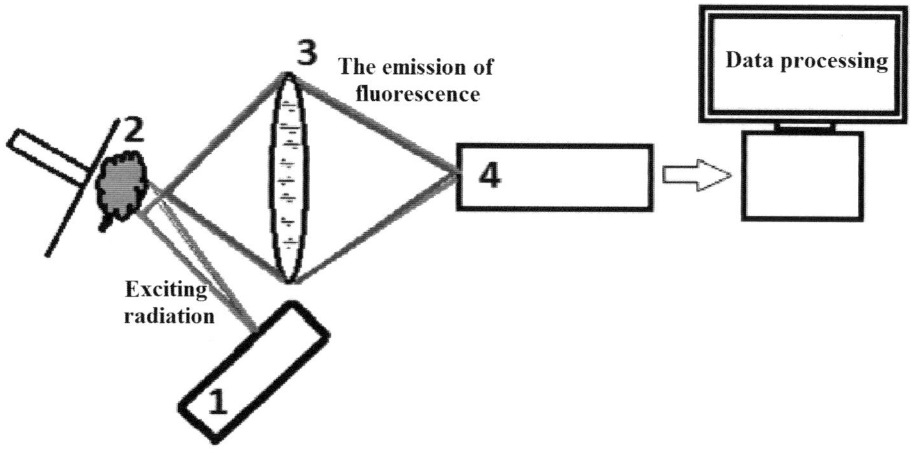

Figure 3. Installation scheme. 1) Light source; 2) Test sample; 3) Focus lens; 4) Spectrofluorimeter.

The installation has the following components:

1) Light sourse - it is intended for irradiation of the sample with light with the necessary characteristics such as: irradiation PPFD (Photosynthetic Photon Flux Density), spectral composition of light, time of illumination. The lighting system provides illumination in several irradiation modes, which are presented in Table 1:

Table 1. Dependence of irradiation intensity of red, white and blue LEDs on current.

Current, A	Voltage, W	Irradiance, W/m^2 (red)	Irradiance, W/m^2 (blue)	Irradiance, W/m^2 (green)	PPFD mol $m^{-2}s^{-1}$ (red)	PPFD mol $m^{-2}s^{-1}$ (blue)	PPFD mol $m^{-2}s^{-1}$ (green)
0.1	9.23	3.94	2.56	1.19	21.4	9.8	5.5
0.2	9.652	7.92	5	2.4	43	19.1	11.1
0.3	10.036	11.8	7.39	3.56	64.1	28.2	16.5
0.4	10.392	15.5	9.24	4.48	84.1	35.3	20.8
0.5	10.739	19.2	11.5	5.61	104.8	44.1	25.9
0.6	11.062	22.5	13.1	6.39	121.4	49.8	29.5

0.7	11.292	26	14.5	7.11	141.9	55.3	33

2) focusing lens with a focal length of 200mm-collects the rays coming from the test sample in the focal plane the monochromator slit is installed in the focal plane.
3) Monochromator MDR 204-designed for measurements of radiation spectra and kinetic measurements.

As a result of the measurements carried out on this installation according to the luminescence analysis technique, the fluorescence intensity was measured at different levels of irradiation.

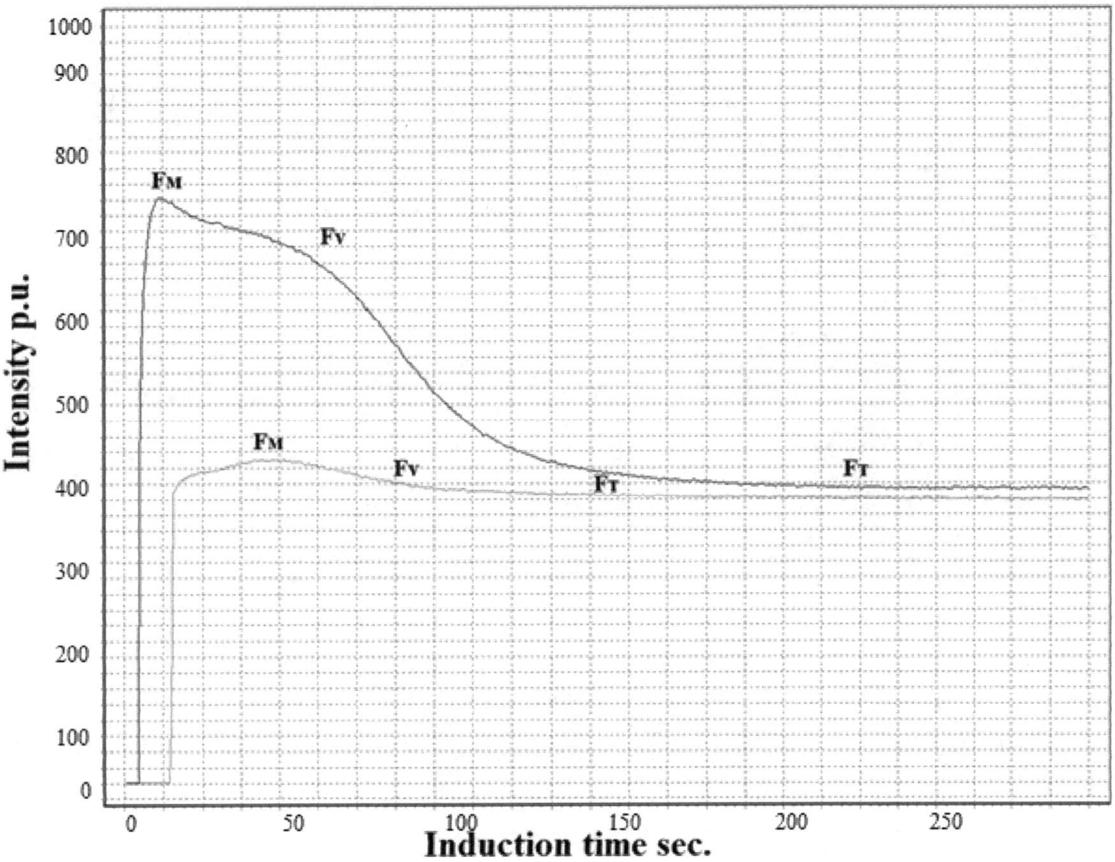

Figure 4. Induction curve.

The chart shows the change in fluorescence intensity over time. At two different levels of learning, characteristic points of the induction curve are obtained, thus the following parameters are determined:

F_M – maximum fluorescence is determined after the adaptation of the plant to the dark with the help of saturating radiation and shows the effectiveness of the initial processes of photosynthesis.

F_V – variable fluorescence is the difference between maximum and initial fluorescence, with the help of this indicator, the efficiency of photosynthesis is determined.

F_T – the steady - state fluorescence level shows the rate of adaptation of the plant to light, which also reflects the efficiency of photosynthesis.

The use of amplitude modulation in the second method expands the possibility of measuring the level of fluorescence, this method gives the opportunity to use of a saturating beam, which causes the processes of extinguishing luminescence, while using a source emitting modulated light, and the sensor in the spectrofluorometer registers only the variable component of radiation [1, 2]. Thus, it is possible to carry out measurements with application of saturating pulses and to investigate how chlorophyll reacts to them.

As part of the work the installation which has which has in its composition LED PPFD light source with adjustable wavelength in the range of 420 to 670 nm was designed. This source was created by saturating pulses with an intensity 1000 - 2000 mol $m^{-2}s^{-1}$ and a spectrofluorometer with amplitude modulation.

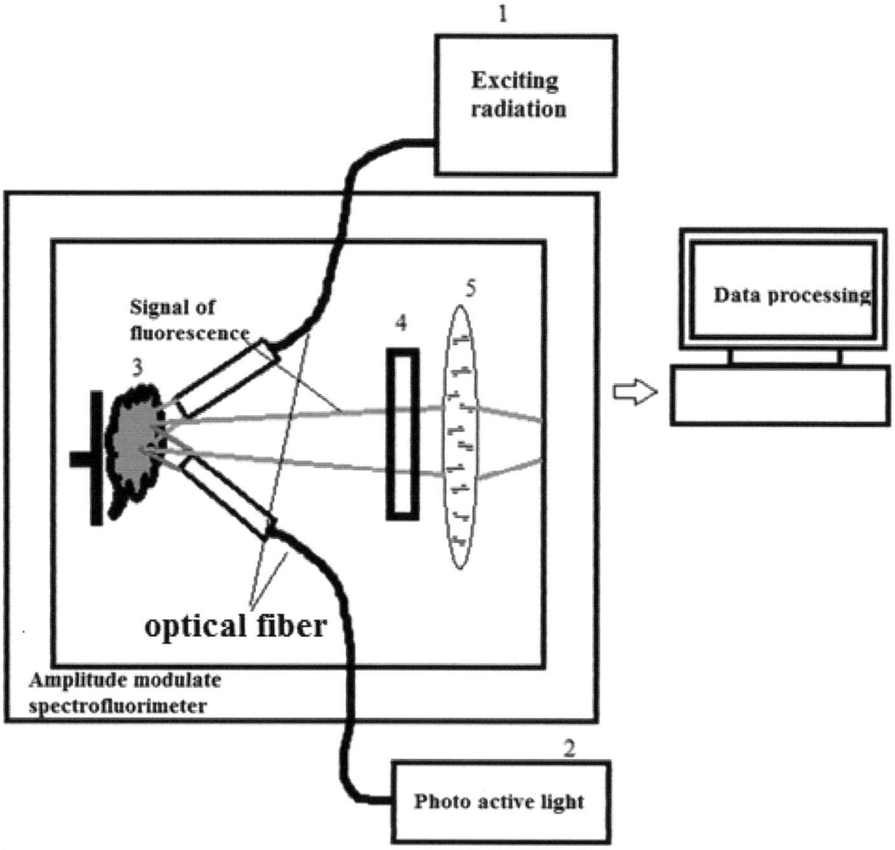

Figure 5. Installation scheme. 1) Exciting radiation source; 2) Photo active light source; 3) Test sample; 4) Light filter of monochromator; 5) Focusing lens.

The chart of induction curve using saturating light.

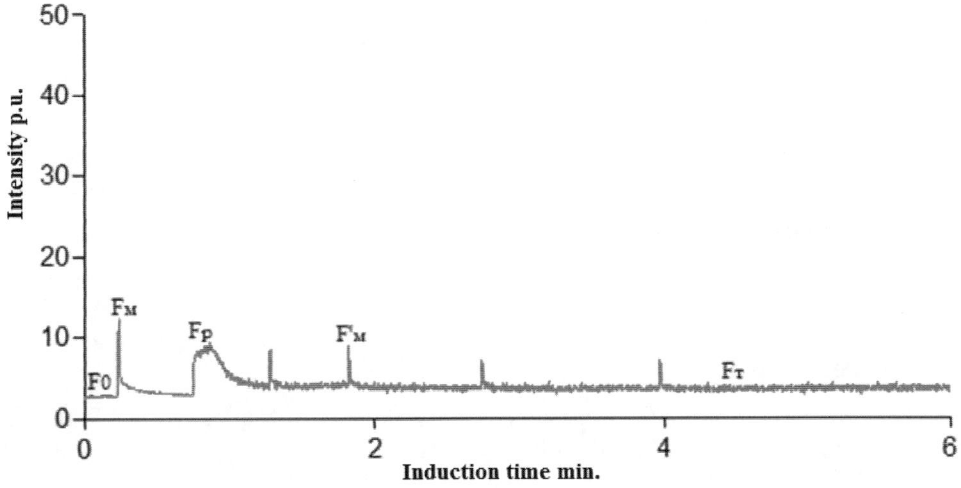

Figure 6. Induction curve with amplitude modulation light.

Max level and start level of luminescence which were obtained from saturating and photoactive radiation (F0, FM, Fp), stationary fluorescence (F_T) [2]. According to this schedule, one can assess the state of the plant under study. Because the level F'_M below the maximum of chlorophyll fluorescence F_M since part of the fluorescence energy is absorbed as a result of extinguishing processes of petrochemical nature, such as heat. This suggests that the plant does not experience stress associated with lack of water or sol balance. F_M – the maximum fluorescence is much higher than the initial fluorescence F_0, which indicates the effective photosynthesis of the plant. As a rule, this parameter is ten times higher than the start fluorescence, as it is seen in the chart.

3. Conclusion

Analysis of chlorophyll fluorescence gives a wide range of possibilities in the selection of the necessary lighting parameters. Such methods allow us to track the reaction of chlorophyll to changes in external conditions, which will contribute in the future to the creation of adaptive irradiation facilities and as a consequence will increase the efficiency of greenhouse complexes.

As part of the work, two research facilities were designed, which were adapted to the existing fluorescence analysis technique. These settings will help to accumulate statistical data on the reaction of chlorophyll to external lighting and allows you to link changes in the intensity of fluorescence with lighting parameters.

Acknowledgement

The work was supported by the Ministry of Education and Science of the Russian Federation under the State Program in the field of scientific investigations №13.3647.2017/PP

References

[1] Goltseva V N, Kalaji H M, Paunova M, Babas V, Khorachek T, Moisky Ya, Kotsel H and Allakhverdiev S I 2016 The use of variable chlorophyll fluorescence to assess the physiological state of the photosynthetic apparatus of plants, *Russian Journal of Plant Physiology* **63 6** 869-893

[2] Rubin A B 2000 Biophysical methods in environmental monitoring. Moscow: Moscow state University. M. V. Lomonosov [in Russian]

IOP Conf. Series: Materials Science and Engineering **754** (2020) 012005 doi:10.1088/1757-899X/754/1/012005

The influence of diverse optical wavelengths on growth rate of Chlorella Vulgaris microalgae

O A Trofimchuk[1], A N Yakovlev[1], S B Turanov[1], M V Trigub[2] and S A Romanenko[1]

[1] Division for Material Science, School of Advanced Manufacturing Technologies, National Research Tomsk Polytechnic University, 30 Lenin Avenue, Tomsk 634050, Russia
[2] V.E. Zuev Institute of Atmospheric Optics SB RAS, 1 Academician Zuev Square, Tomsk 634055, Russia

E-mail: oat1@tpu.ru

Abstract. The effect of different optical wavelengths on growth and development of microalgae had been examined repeatedly. Many articles had been published over the subject, although the results of these investigations were frequently controversial. Depending on the number of days spent on cultivation process, yield gain could be both significant and barely noticeable. Also, irradiation intensity plays vital part in microalgae functioning.

1. Introduction

Microalgae grow and develop in the process of photosynthesis, like other plants. In other words, they assimilate inorganic carbon to convert it into organic constituents. Stimulation of this reaction occurs with the help of light, as a source of energy. Based on this information, the spectral composition and energy component of light should be taken into account when cultivating microalgae. It was revealed that microalgae consume optical radiation in the range from 400 to 700 nm for the photosynthesis process. The wavelengths of light absorbed by microalgae cells differ in groups. Green microalgae, for example, is fed with light energy for photosynthesis, using chlorophylls as the main pigments that absorb light energy in the ranges 450-475 nm and 630-675 nm, and carotenoids as secondary ones - absorption occurs only in the wavelength range 400-550 nm [1]. Various research mentions the yield gain of microalgae under different wavelengths conditions. Red (600-700 nm) and blue (400-500 nm) light stimulate microalgae development, when the rate of growth and lipid content vary due to diverse light intensity [1-3].

2. Materials and methods

In their work *Blair M. F. et al* had proven that adequately selected optical wavelengths affect remarkably yield gain of microalgae. Red and green lights did not cause any positive effect on growth rate in comparison to wavelengths from white and blue bands (figure 1). Compact fluorescent lamps with 60 W power level (276 μmol/m^{2}s) were applied for suspension irradiating. To avoid thermal effect, light sources were located 20 cm apart from the bioreactors.

Figure 1 b illustrates the highest growth rate under white, blue and red lights on the 3rd day of cultivation: 0.369 d^{-1}, 0.235 d^{-1} and 0.140 d-^{1} respectively. Green lighting has the highest growth rate on the 2nd day: 0.137 d^{-1}. Green-band photons have 20% more energy than red photons (680 nm) and 15.5% less than blue ones (470 nm). Therefore, green light could have decelerating effect, caused by power of provided energy. However, it is obvious, that blue photon has more energy than green. It is depicted in the figure 1 c that the highest biomass productivity was 0.038 g/L-d, 0.0199 g/L-d and 0.0096 g/L-d for white, blue and red lights on the 3rd day respectively. For green-light exposure this criterion was far less – 0.0088 g/L-d on the 2nd day.

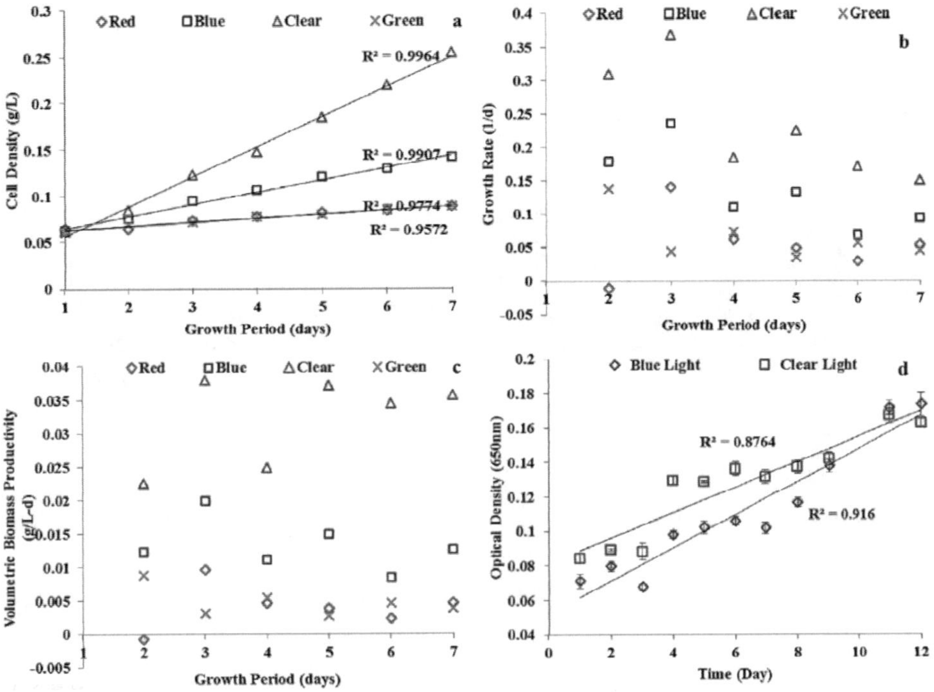

Figure 1. The influence of optical wavelength on: (a) cell density; (b) growth rate; (c) volume biomass productivity; (d) optical density during 12-day period.

It appeared that 7-day results shouldn't be represented as complete growth cycle of microalgae, apart from red and green lights, which sequentially demonstrate lower growth rates and mismatch practical appliance as provided by the results of the experiment. Another set of experiments was conducted with the use of blue and white lights for 12 days. Figure 1 d depicts that growth rate of microalgae under blue light conditions surpassed growth rate with white lighting (after 10 days). This confirms the intensification of microalgae Chlorella growth rate under blue light after the adaptation period [1].

In that investigation red light did not demonstrate high growth rate. This tendency can be explained with the help of previous research group that claimed that red light could actually cause damage in cells [3, 4]. Moreover, it is noteworthy, that specific ability of optical wavelengths absorption depends on the microalgae species. During the experiment, Yan et al had detected that white and red lights performed with far higher yield gain of microalgae, although the study was conducted between 6th and 10th days and the cultivation medium was synthetic high-tension wastewater [2]. In addition, growth properties were affected by light intensity [5].

Figure 1 d shows the approximate linearity of growth rate under blue-light irradiation ($R^2 = 0.916$), in contrast to white-light irradiation ($R^2 = 0.8764$), or highlights the trend towards exponential phase reaching. Cultivation under blue-light irradiation demanded serious time input for exponential phase achieving. In this case, slow growth rate under blue light does not satisfy the requirements for practical use, because the more rapid is the exponential phase reaching, the more efficient would be biomass production and harvesting time would be less [1].

In another research by *Barghbani R. et al* more yielding microalgae cultivation was obtained under blue and yellow lights. The results from the study above match with the conclusions of other scientists. Throughout the actual testing the irradiance was continuous, the light sources were luminescent lamps of different colors and power level was 23 W (figure 2).

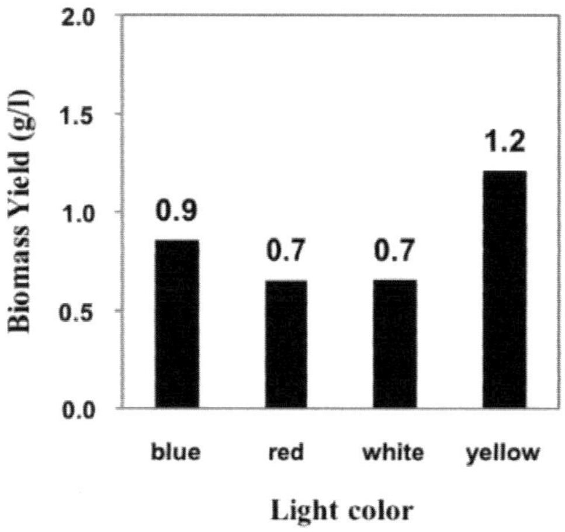

Figure 2. The influence of different colors on extracted biomass yield (dry basis) during 7-day cultivation in various experimental conditions according to Taguchi.

Four diverse irradiation colors were used in this examination (white, yellow, blue and red) (figure 2). Microalgae demonstrated peak growth rate under yellow-light impact (low energy). Other colors (blue, white and red) led to slightly less biomass productivity [6].

In the following research by *Kim D. G. et al*, continuous irradiation was introduced at the light intensity level of 100 μmol/m^{2}s under monochromatic blue (λ_{max} = 430-465 nm) and red (λ_{max} = 630-665 nm) lights. It was illustrated that particular wavelength could dramatically affect the growth rate of microalgae (figure 3).

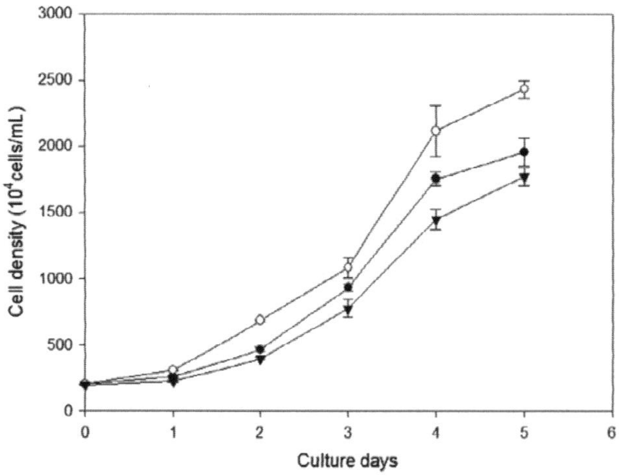

Figure 3. The influence of three varied wavelength bands on Chlorella yield gain, where red (-○-), white (-●-) and blue (-▼-) the relation between cell density and days of cultivation.

Cells number depended remarkably on wavelength of light due to the fact that Chlorella microalgae cumulated the highest and the lowest cells density under red and blue light relatively. Cells density of Chlorella grown under red light was 1.5 times higher than density of cells, irradiated by blue light during post-inoculation time period (figure 3). These results correlate with the conclusion of scientists, who conducted similar experiments with other microalgae species. Red-band wavelengths appeared the most optimal for effective Spirulina platensis growth as well as for Chlorella pyrenoidosa [8]. Pigmentation enhancement of chlorophyll was initiated by red light, what could be referred to the positive effect of this exact irradiation [9]. Yet white-light irradiation of suspension led to average cells number increase, while blue light gradually was dwindling cells gain. One of the possible assumptions is that an excess of light wavelengths from the blue spectrum can inhibit the division of microalgae cells, which leads to a decrease in the growth of microalgae biomass.

3. Results

It is indicative that cytomorphology of sizes was significantly modified as a function of light wavelength. Red light multiplied the number of cell divisions, causing small-sized cells appearance, and at the same time blue light provided size development of cells. On that basis, it could be assumed that microalgae cells generation with larger dimensions at the initiation of cultivation could encourage further cell fission at advanced development stage, and biomass productiveness raise consequently [7].

During microalgae stock culture transfer into photobioreactor adaptation to ambient conditions occurs at first, and cells growth prediction becomes insufficient. Also, to obtain valid results, microalgae should be cultivated before cells growth rate reaches stationary phase. Under different wavelengths and irradiation level conditions stationary response time varies for all samples.

4. Conclusion

Nowadays, microalgae are widely applied in diverse branches of human activity such as pharmaceutics, wastewater treatment, agriculture, biodiesel fuel production et al. To achieve accelerated growth of microalgae, vast number of parameters should be taken into consideration, particularly, optimum spectral range selection. Artificial lighting sources have advantage over natural – the possibility to control spectral structure which enables choosing the most efficient light conditions while cultivating. In this case, there is a possibility to obtain high rates of valuable substances, which are included into Chlorella, biosynthesis, and enhance reproduction and yield gain of the culture. Cultivation time of microalgae could be noticeably reduced, if chosen spectral structure of irradiation sources would correspond to absorption spectrum of Chlorella.

Acknowledgments

The work was supported by the Ministry of Education and Science of the Russian Federation under the State Program in the field of scientific investigations №13.3647.2017/PC.

References

[1] Blair M F, Kokabian B, Gude V G 2014 Light and growth medium effect on Chlorella vulgaris biomass production *J. of Envir. Chem. Engin.* **2** pp 665-674

[2] Yan C, Zhang L, Luo X, Zheng Z 2013 Effects of various LED light wavelengths and intensities on the performance of purifying synthetic domestic sewage by microalgae at different influent C/N ratios *Ecol. Eng.* **51** pp 24-32

[3] Khalili A, Najafpour G D, Amini G, Samkhaniyani F 2015 Influence of Nutrients and LED Light Intensities on Biomass Production of Microalgae Chlorella vulgaris *Biotech. and Biopro. Eng.* **20** pp 284-290

[4] Ruyters G (1984) Effects of Blue Light on Enzymes. In: Senger H. (eds) Blue Light Effects in Biological Systems *Proceedings in Life Sciences* pp 283-301

[5] Khoeyi Z A, Seyfabadi J, Ramezanpour Z 2012 Effect of light intensity and photoperiod on biomass and fatty acid composition of the microalgae, Chlorella vulgaris *Aquacult. Int.* **20** pp 41-49

[6] Barghbani R, Rezaei K, Javanshir A 2012 Investigating the effects of several parameters on the growth of Chlorella vulgaris using Taguchi's experimental approach *Int. J. Biotechnol. Wellness Ind.* **1** pp 128-133

[7] Kim D G., Lee C, Park S M, Choi Y E 2014 Manipulation of light wavelength at appropriate growth stage to enhance biomass productivity and fatty acid methyl ester yield using Chlorella vulgaris *Biores. Tech.* **159** pp 240-248

[8] Wang C Y, Fu C C, Liu Y C 2007 Effects of using light emitting diodes on the cultivation of Spirulina platensis *Biochem. Eng. J.* **37** pp 21-25

[9] Matthijs H C, Balke H, Van Hes U M, Kroon B, Mur L R, Binot R A 1996 Application of light emitting diodes in bioreactors: Flashing light effects and energy economy in algal culture (Chlorella pyrenoidosa) *Biotechnol. Bioeng* **50** pp 98-107

IOP Conf. Series: Materials Science and Engineering **754** (2020) 012006 doi:10.1088/1757-899X/754/1/012006

Threshold characteristics of non-resonator laser generation in dye solutions with ZnO nanoparticles

Al A Zemlyanov[1,2], A V Trifonova[2] and R V Ryambov[2]

[1] V. E. Zuev Institute of Atmospheric Optics of the Siberian Branch of the Russian Academy of Sciences, Tomsk, Russia
[2] National Research Tomsk State University, Tomsk, Russia

E-mail: zeml1616@mail.ru

Abstract. An experiment was carried out to determine the generation thresholds in solutions of the Rhodamine 6G dye with Au nanoparticles having a plasmon resonance at the pump wavelength and ZnO nanoparticles not having a plasmon resonance. It was shown that solutions with ZnO nanoparticles have lower lasing thresholds in comparison with solutions with Au nanoparticles.

1. Introduction

The development of laser emitters in recent years is associated with their significant miniaturization and the creation of highly efficient active media. One of these lasers is considered to be a stochastic laser, known in the foreign literature as random-laser [1-3]. The active medium of this laser is a composite of laser-active molecules of nanoparticles of various materials.

In a classical laser, generation is usually determined by two factors - optical amplification, which is created by the inversion of the population in the excited active substance, and the formation of feedback by the laser resonator. In this case, the phenomenon of light scattering plays a negative role in this situation and is always minimized. However, in a stochastic laser, a chaotic strongly scattering medium improves the conditions for obtaining a laser effect in it.

In [1], the author H. Cao defines a stochastic laser as a resonatorless laser (due to the absence of external mirrors), in which the positive feedback mechanism is realized due to light scattering in an active medium with optical inhomogeneities.

According to the type of stimulated emission, random lasers can be divided into two classes [1]. The first of which are random lasers based on solutions of laser dyes with embedded nanoparticles of various substances. In such structures, the effect of multiple scattering by small inhomogeneities increases the residence time of stimulated emission photons in an excited active medium. The second class includes stochastic lasers, the principle of generation of which is based on the appearance of stimulated emission in the so-called micro-resonators, the mirrors of which are well-reflecting faces of microcrystals. Closed resonators are formed randomly during the action of the pump pulse.

At present, it is generally accepted that resonance-free laser generation in a scattering amplifying medium is realized either due to an increase in the interaction time of secondary radiation photons with active substance molecules, or due to the plasmon resonance effect. However, from previous experiments in [4], it was shown that the use of plasmon resonance nanoparticles does not lead to a significant decrease in the lasing thresholds.

A stochastic laser has special properties with respect to other types of lasers, namely a simple production technology, a wide angular dependence and the possibility of tuning the wavelength in a wide spectral range. The angular distribution of the output signal of a random laser is very wide and distributed at a total solid angle of 4π. Such wide angular radiation is, in principle, ideal for use in creating displays. A particular advantage of the active media of random lasers is that they can be prepared in the form of suspensions, which can be used as coatings on surfaces of arbitrary shape.

This communication demonstrates a comparison of the generation thresholds in Rhodamine 6G solutions with Au nanoparticles with plasmon resonance at the pump wavelength ($\lambda = 532$ nm) and ZnO nanoparticles that do not have this effect.

2. Experimental setup

The experiment was conducted on the setup shown in Figure 1.

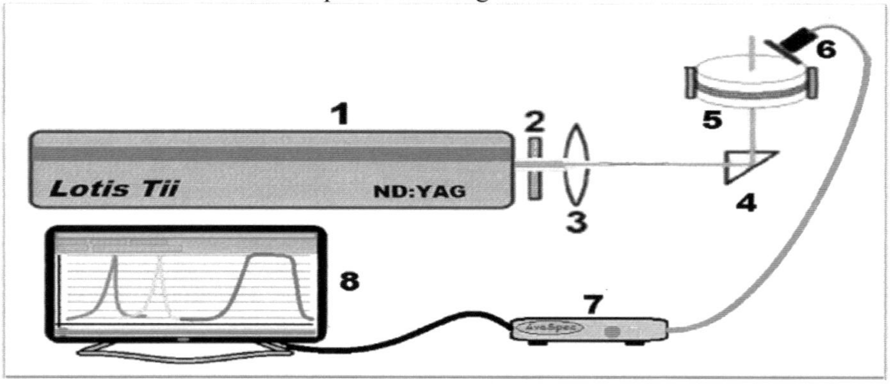

Figure 1. Scheme of the experimental setup. 1 - laser, 2 - light filter, 3 - focusing lens, 4 - prism of total internal reflection, 5 - cuvette, 6 - input window of the receiving system, 7 - spectrometer, 8 - PC.

During the experiments, the studied solution with a thickness of 1 mm was placed in a cuvette onto which the radiation from an aluminum – sodium garnet laser was incident with a pulse duration of 6 ns and a wavelength of 532 nm. Then, radiation passing through a light filter collecting a lens and a prism of total internal reflection fell on a cell with a working solution. The luminescence spectra of the solution were recorded using an Avaspec spectrometer. The obtained spectra were displayed on a PC.

3. Results

Typical spectra of spontaneous luminescence and stochastic generation are presented in Figure 2.

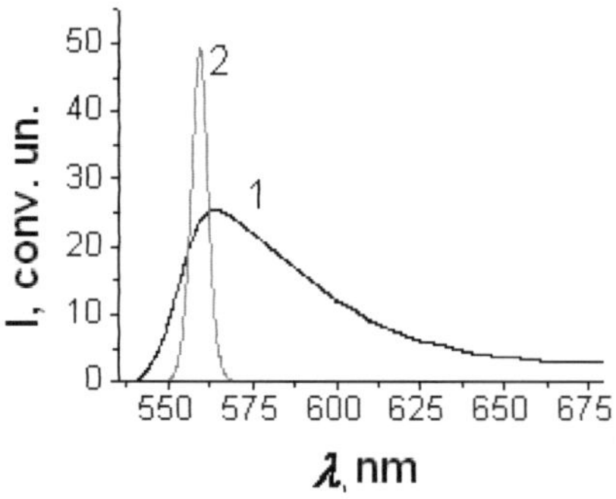

Figure 2. Typical spectra of spontaneous luminescence (1) and stochastic generation (2).

From the spectra, we plotted the dependences of the half-width of the spectrum and the luminous intensity on the laser pump energy. Dependencies are presented in figure 3.

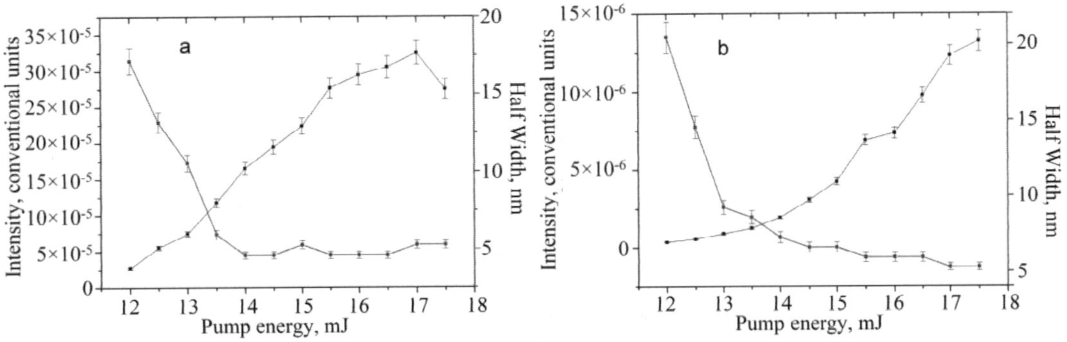

Figure 3. The dependence of the half-width of the spectrum and the intensity of the luminescence on the pump energy for solutions with Au nanoparticles (a) and ZnO (b) with a concentration of nanoparticles in solution - 1.459×10^{-6}% volume fractions.

Using the emission spectra and dependences presented in Figure 3, we determined the generation thresholds in solutions with Au and ZnO nanoparticles at various nanoparticle concentrations. The dependences of the threshold values on the nanoparticle concentrations are presented in Figure 4.

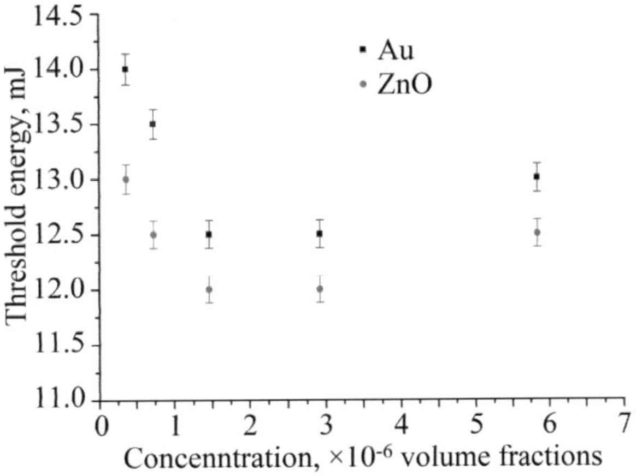

Figure 4. Dependence of threshold energy on the concentration of nanoparticles.

As can be seen from the dependences, solutions with ZnO nanoparticles at all used concentrations have lower lasing thresholds in comparison with solutions with Au nanoparticles.

4. Summary

As a result of the experiment, generation thresholds in solutions with Au and ZnO nanoparticles were obtained. The results indicate that the plasmon resonance possessed by Au nanoparticles at the used pump wavelength does not in any way affect the lowering of the generation thresholds. Such conclusions can be made based on the fact that the zinc oxide nanoparticles used in the experiment cannot possess plasmon resonance, since they are dielectric. Moreover, the generation thresholds in solutions with ZnO are equal to or less than the thresholds in solutions with Au nanoparticles.

References

[1] Cao H 2003 *Topical Review* V **13** P 1-39
[2] Lawandy N M, Balachandran R M. 1995 *Nature (London)* **373** 204-208
[3] Balachandran R M, Pacheco D P, Lawandy N M 1996 *Applied Optics* **35**, 4 640-646
[4] Donchenko V A, Zemlyanov Al A, Panamarev N S et.al. 2018 *Proc. Of SPIE* **10833** 108332C-1–4

IOP Conf. Series: Materials Science and Engineering **754** (2020) 012007 doi:10.1088/1757-899X/754/1/012007

Research of surface microstructure of the steel 40CrNiMo after EDM

A Grisharin, T Ablyaz, K Muratov, E Shlykov, E Morozov, A Shafranov, E Matigullina and L Sirotenko

Perm National Research Polytechnic University, Perm-614000, Russian Federation

*E-mail: lowrider11-13-11@mail.ru

Abstract. The aim of the work is to study the influence of modes of electrical discharge machining, on the microstructure and microhardness of the treated surface of thermally hardened steel 40CrNiMo. To fulfill the task, a sample was made of hardened steel 40CrNiMo. The sample was processed in various modes of electrical discharge machining. Microsections were made from the obtained sample. Metallographic and microdurometric analysis was carried out. The study of the microstructure showed that no clearly defined altered layer was detected in all processing modes. The study of microhardness showed that near the surface of the studied samples there is a drop in the level of microhardness. This phenomenon is caused by tempering during the surface heating during processing. The heat-affected zone does not exceed 1.2 mm, and is minimal under high-performance modes.

1. Introduction

In the modern world, the competitiveness of industrial enterprises is determined by the flexibility of production. First of all, this is achieved by introducing new technologies and processing methods. For the manufacture of spatially complex parts by traditional machining, special technological equipment and expensive cutting tools are required. This approach makes the process of readjusting production time-consuming and highly costly. The use of alternative methods for manufacturing parts by three-dimensional copying, such as EDM, is the most promising [1].

The essence of the electric discharge machining method is the gradual melting of particles of the processed material under the action of current pulses, followed by leaching of the resulting erosion products due to the hydrodynamic forces of the working fluid [2]. This method provides the ability to process materials regardless of their hardness. EDM allows obtaining high geometry accuracy and surface finish [3]. Currently, the method of EDM is widely used for the manufacture of work surfaces of the main parts of machines and mechanisms directly involved in the kinematics of work.

A characteristic phenomenon in EDM is the appearance on the treated surface of an altered surface layer. The composition and properties of the changed layer to a greater extent depend on the modes of electrical discharge machining, the processed material, the material of the electrode-tool, as well as on the composition of the working fluid and the process conditions [4]. Since the properties of the modified surface layer can differ significantly from the properties of the base material, this can adversely affect the operational properties of the parts. The study of the influence of modes of electric discharge machining on the structure and properties of the treated surface is important.

For trouble-free operation of product mechanisms, it is necessary to ensure high reliability of the main parts involved in the kinematics of the nodes. This is achieved by using wear-resistant steel grades for their manufacture. Steel should have a high complex of mechanical properties, providing high resistance to wear in the range of operating temperatures [5, 6]. For these purposes, alloyed thermally hardened steels are widely used, including 40CrNiMo steel.

The process of electric discharge machining of alloy steels is accompanied by a change in the structure of the surface layer consisting of several zones, Figure 1.

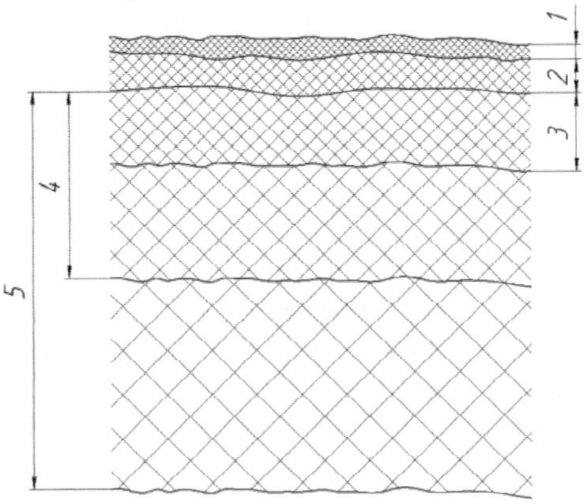

Figure 1. The surface layer of alloy steel after EDM.
1. The zone of saturation with elements of the working fluid; 2. The deposition zone of the material of the electrode-tool; 3. White layer; 4. The zone of thermal influence; 5. The zone of plastic deformation.

The white layer consists of the molten material of the workpiece, the material of the electrode tool and the products of the pyrolysis of the working fluid. The thickness of the white layer depends on the EDM modes, as well as on the thermal conductivity of the material of the workpiece. When processing steels under mild processing conditions, the white layer is distributed on the treated surface in separate sections, up to 40 microns thick and 250-260 microns long. The size of the white layer in the processing of preheated workpieces in soft conditions is little dependent on the temperature of the workpieces. In rough processing conditions, the white layer completely covers the treated surface, and the maximum layer thickness is 165-310 microns. Under the white layer is a phase transition zone caused by heating during erosion. Depending on the material, this zone may contain microcracks, residual stresses and other defects. The zone of phase transformation is followed by the zone of altered grain structure or the released zone [4]. The surface layer of alloyed steels after electric discharge machining at high pulse energies is qualitatively and quantitatively different from the surface layer after processing at minimum conditions [7].

Overheating of the processed material leads to re-tearing and tempering of the material in the processing zone. In this case, in addition to changing the physicomechanical properties of the material, internal stresses occur, which, in turn, can lead to premature failure of the part during operation, and even in the manufacturing process [8].

An analysis of literary sources showed that, due to the lack of understanding of the issue, predicting the structure and surface properties of structural alloyed heat-strengthened steels after electric discharge machining is difficult depending on the processing regimes, there are no recommendations on the designation of processing modes that provide minimal changes in the surface layer.

The purpose of work is to study the influence of modes of electric discharge machining on the structure and properties of structural heat-strengthened steels, not an example of 40CrNiMo steel.

2. Materials and methods

The studies were carried out on a heat-treated (quenching in oil - 850 °C followed by an average tempering of 450 °C in air) sample of 40CrNiMo steel having, after heat treatment, the structure of tempering troostite. EDM processing was carried out by copper ET on a Smart CNC machine at reference modes in accordance with Table 1.

Table 1. Parameters of the EDM modes of a sample of steel 40CrNiMo.

№	I, A	Ton, us	U, V
1	2	40	50
2	2	150	50
3	8	40	100
4	8	150	100

The microstructure was studied on microsections using an OLYMPUS GX 51 light microscope at magnifications up to 1000. Image processing was performed using OLYMPUS Stream Motion software.

Sections were made by pressing samples in a metkon ECOPRESS 100 automatic mounting press and then grinding them on sandpaper with grain sizes from r240 to r1500. After that, polishing was carried out on the canvas using diamond paste. To reveal the structure, the microsection was etched with a 4% solution of nitric HNO3 acid in ethanol.

Microdurometric tests were performed on a PMT-3 microhardness tester. The hardness on the PMT-3 microhardness meter was determined by the method of the reconstructed print by indentation of a tetrahedral diamond pyramid with a square base at a load of 25 grams with a holding time of 10 s.

3. Results

After the implementation of the planned modes of electrical discharge machining, the test samples were cut in cross section using wire cutting EDM machine EcoCut. Next, microsections were prepared in accordance with the claimed methodology, on which microhardness paths were applied with a PMT-3 device from the treated surface deep into the core of the sample. A cross-sectional microstructure was also examined using an Olympus GX-51 microscope. The microhardness measurement results are shown in Figures 2-5.

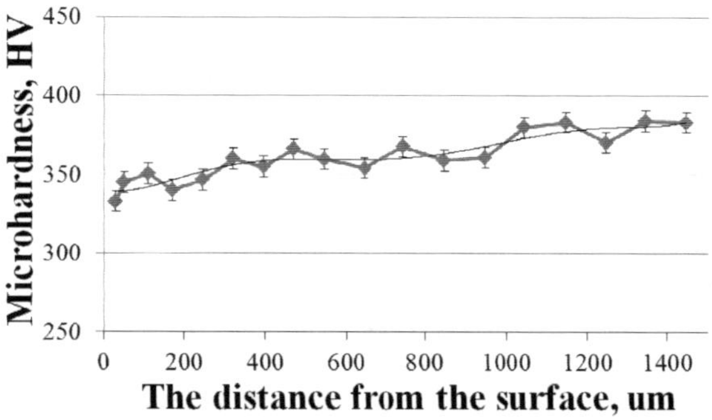

Figure 2. Dependence of microhardness along the depth of steel 40CrNiMo from the surface to the core after EDM according to mode No. 1

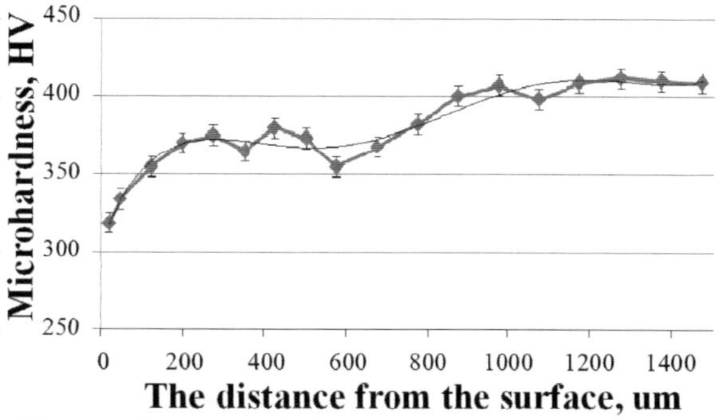

Figure 3. Dependence of microhardness along the depth of steel 40CrNiMo from the surface to the core after EDM according to mode No. 2

IOP Conf. Series: Materials Science and Engineering **754** (2020) 012007 doi:10.1088/1757-899X/754/1/012007

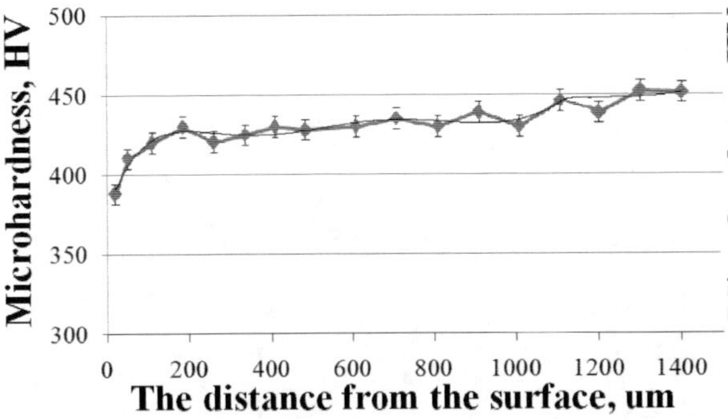

Figure 4. Dependence of microhardness along the depth of steel 40CrNiMo from the surface to the core after EDM according to mode No.3

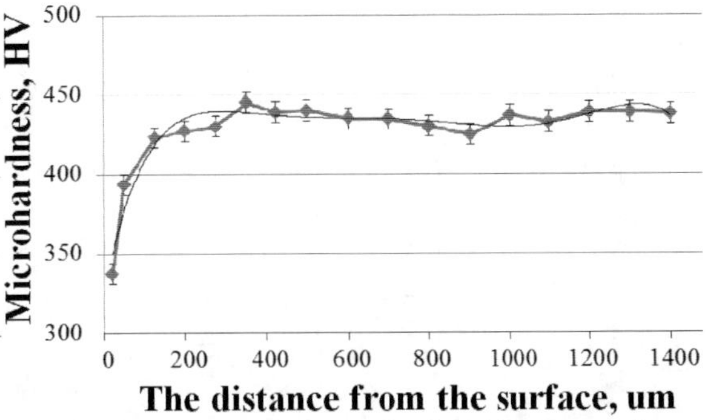

Figure 5. Dependence of microhardness along the depth of steel 40CrNiMo from the surface to the core after EDM according to mode No. 4

A study of the depth distribution of microhardness under various modes of electric discharge machining (modes No. 1 and No. 3 are low-power, and modes No. 6 and No. 8 are powerful) showed that a decrease in the microhardness level is observed near the surface of the samples under study, which is apparently caused by heating the surface of the samples in the process of electrical discharge machining. The process of additional release of the original structure has occurred. This heat affected zone does not exceed 1.2 mm. Moreover, a preliminary analysis showed that when working in high-performance, powerful modes, a minimal zone of thermal influence is observed, which is caused by a shorter exposure time of the heating to the part due to greater productivity, as well as greater erosion of the part.
The results of the study of the microstructure of the samples after EDM are presented in Figures 6-9.

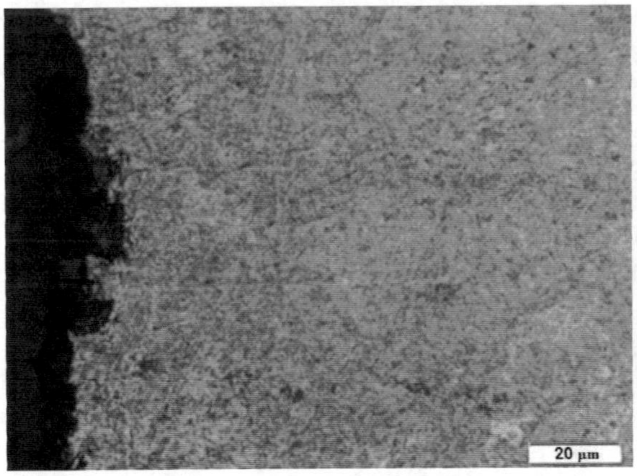

Figure 6. The microstructure of the surface of steel 40CrNiMo after electrical discharge machining mode No.1

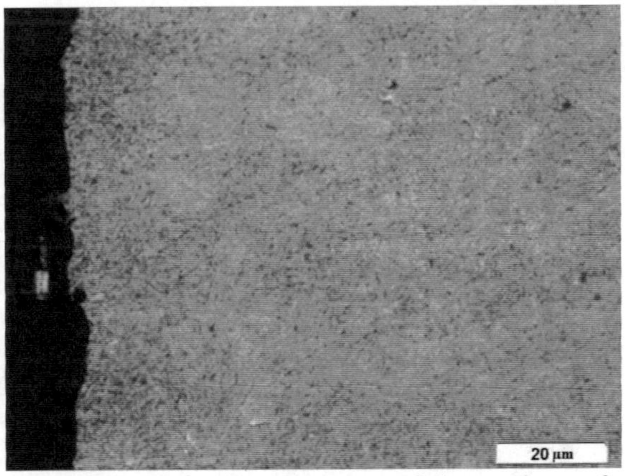

Figure 7. The microstructure of the surface of steel 40CrNiMo after electrical discharge machining mode No.2

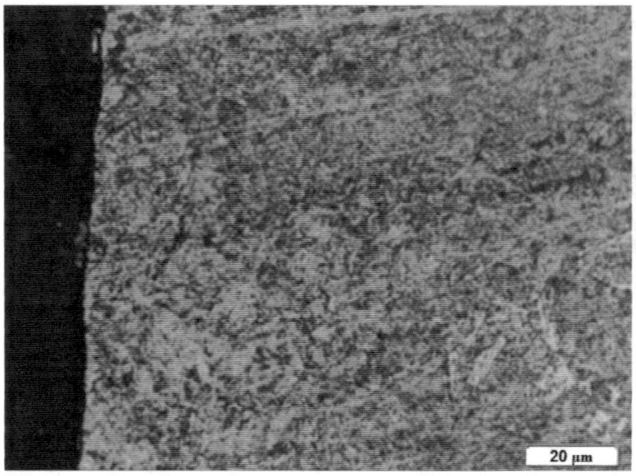

Figure 8. The microstructure of the surface of steel 40CrNiMo after electrical discharge machining mode No.3

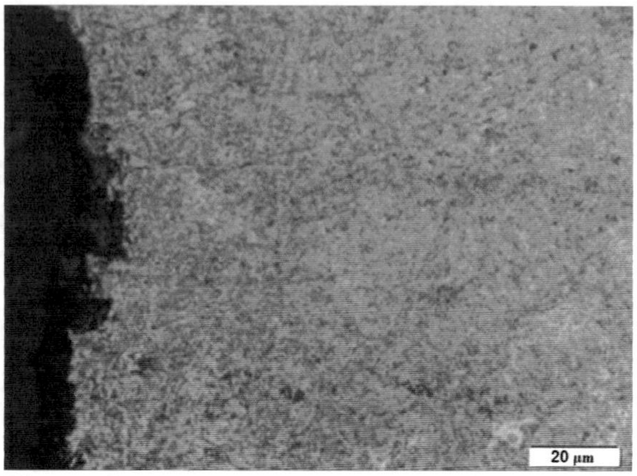

Figure 9. The microstructure of the surface of steel 40CrNiMo after electrical discharge machining mode No.4

The initial structure of the studied steel is a dispersed tempering troostite with a hardness of 440-460 HV. Under all processing modes by metallographic analysis methods, a clearly defined so-called “white layer” was not revealed. This indicates that there is no overheating of a significant part of the surface to temperatures above the melting point and kneading of the material of the electrode-tool, synthetic oil and electrode-parts. Dispersed tempering structures are observed throughout the depth of 40CrNiMo steel.

4. Conclusions

Based on the results of the experiment and analysis of the data obtained we can draw the following conclusions:

1. Metallographic studies of the surface of 40CrNiMo steel after various modes of electrical discharge machining showed that the “white layer” is not detected using light microscopy methods.
2. An analysis of the distribution of microhardness in depth from the surface to the depth of the 40CrNiMo steel under study after various modes of electrical discharge machining showed the presence of a heat-affected zone, which is characterized by a drop in hardness at the surface of the samples as a result of additional tempering under the influence of heating during the action of the electrode tool.

Acknowledgement
Financial support was provided by the Russian Ministry of Education and Science (project 11.9716.2017/8.9).

References

[1] Bains P S, Sidhu S S, Payal H S 2016 *Fabrication and machining of metal matrix composites* (Kapurthala: IKG Punjab Technical University) p 533
[2] Smolentsev V 1983 *EDM and electrochemical methods of processing materials* vol 2 (Moscow: Higher School) p 247
[3] Yeliseyev Y, Savushkin B 2010 *EDM of aerospace products* (Moscow: MSTU him NE. Bauman) p 437
[4] Ploshkin V 2016 *Structural and phase transformations in the surface layers of steels during electrical discharge machining*. (Moscow: Diss. ... doc. tech. Sciences: 05.16.01) p 281
[5] Grigorenko V 2014 *Features of the use of corrosion-resistant steels*. (Moscow: STEEL No. 1) p 60
[6] Melnikov A 2012 *Material Science, Lecture notes Part 2* (Samara: The Ministry of Education and Science of Russia) p 83
[7] Ablyaz T, Panov D, Abrosimova A 2013 *Metallographic analysis of the surface of 65G steel*

after electrical discharge machining (Perm: Modern problems of science and education, No. 5) access mode to the journal: http://www.science-education.ru/.
[8] Amorim F, Weingaertner W 2004 *Die-sinking electrical discharge machining of a high-strength copper-based alloy for injection molds* (Congresso Brasileiro de Engenharia de Fabricação) p 137

IOP Conf. Series: Materials Science and Engineering **754** (2020) 012008 doi:10.1088/1757-899X/754/1/012008

Design of tools with the cutting part of the original profile for high-speed milling

M S Vakulin, Yu I Gordeev and V B Yasinsky
Siberian Federal University, 79, Svobodny pr., 660041 Krasnoyarsk, Russia

E-mail: tms-mtf@rambler.ru

Abstract. This paper considers and justifies the design of the cutting part of multi-edge contour mills with the increased performance and surface quality. The numerical experiments using the finite element method (FEM) allow determining stresses and deformations in the layer of the material being cut when processed by multi-edge mills of a new type, i.e. they allow evaluating indirectly unit loads during milling. The required shape and dimensions of the cutting wedge were determined taking into account various values of the geometric parameters of the cutting part, the properties of the workpiece material and cutting conditions. All the above resulted in a 3D model of the end mill, which had got a trapeze-shaped tooth located along two intersecting helical lines with tilt angles of 22 and 85 degrees. Along with improving the quality of the processed surface, the experimental studies also showed a change of chip shape. The chip has a finely crushed structure; the sizes of its elements are about 2 μm, which agrees nicely with the results of the FEM preliminary estimates.

1. Introduction

The development and active use of multi-axis machines with a high-speed spindle provide high precision machining and highly efficient manufacturing of complex shapes. It is possible to machine a workpiece surface by controlling the cutting edge geometry and angles of end mill which provide the reducing of the cusp height and improve the surface roughness [1]. Micromilling with high geometrical complexity of tool enables reducing the volume and sizes of chips and the cutting force as well as producing 3D micro-components of a wide variety of metallic and non-metallic materials, most significantly composites. Additionally, turning with the use of a single-point cutting tool is proposed to be substituted with end-mill turning, so-called turn-milling [2]. When the chip volume decreases, the cutting thickness can be similar to the tool edge radius size. This chip formation mechanism causes the so-called size effect – a phenomenon characterized by a substantial increase of the specific cutting force for machining processes at small cutting thicknesses, i.e. a reflection of the increase in a part shear flow stress due to the decreasing cutting zone [3]. The use of the known analytical methods of calculation based on the geometric and stereological models is limited or unsuitable to design high-speed mills because it does not take into account the high intensity and rate of deformation processes. The experimental studies also investigate the size effect and chip formation in microscale cutting [4]. Using the analytical and numerical approaches and the finite element model, we can determine the minimum thickness of the material uncut with the chip, and the geometry of tools [5, 6]. Therefore, the development of new cutter designs and methods of numerical evaluation of the deformation processes in the cut layer based on the finite element method (FEM) is a topical and practically significant task.

The aim of this research is to optimize the geometry of multi-bladed contour mills in order to increase productivity and quality of processing.

2. Mill design

Several types of milling routers by SANDVIK Coromant and ISCAR, which combine the principles of milling and crushing, have been selected as the base objects for a comparative study. Many researchers use specific cutting force when milling regardless of other parameters to evaluate the processing with the use of such mills. This approach is feasible as, despite the same cutting conditions and cutting section, the specific cutting force during micromilling is 90% greater than during macromilling. This fact indicates that when changing the geometry of the cut layer section (i.e. the shape of the cutter tooth) and the shape of the chip, there appear different specific cutting forces. Analyzing the relationship of feeds per tooth and the cutting depth on the specific cutting force, we can conclude that their influence is not significant for micromilling, in contrast to macromilling [2]. Thus, we can consider that specific cutting force is a more important parameter which influences processing and surface quality of a workpiece during high-speed micromilling. It is assumed that decreasing the specific cutting force will improve the efficiency and quality of machining.

The first stage of the study was an efficiency analysis of new end mill geometry with the increased number of cutting edges which was carried out with the use of common models. We optimized the geometric parameters of the cutting part and developed a 3D model of the mill (figure 1) based on the conducted numerical experiments. This new design is supposed to have an increased number of cutting cycles in comparison with its counterparts, which means that the specific cutting force on each tooth edge will be minimized, and the profile of a surface being processed should have a reduced roughness value.

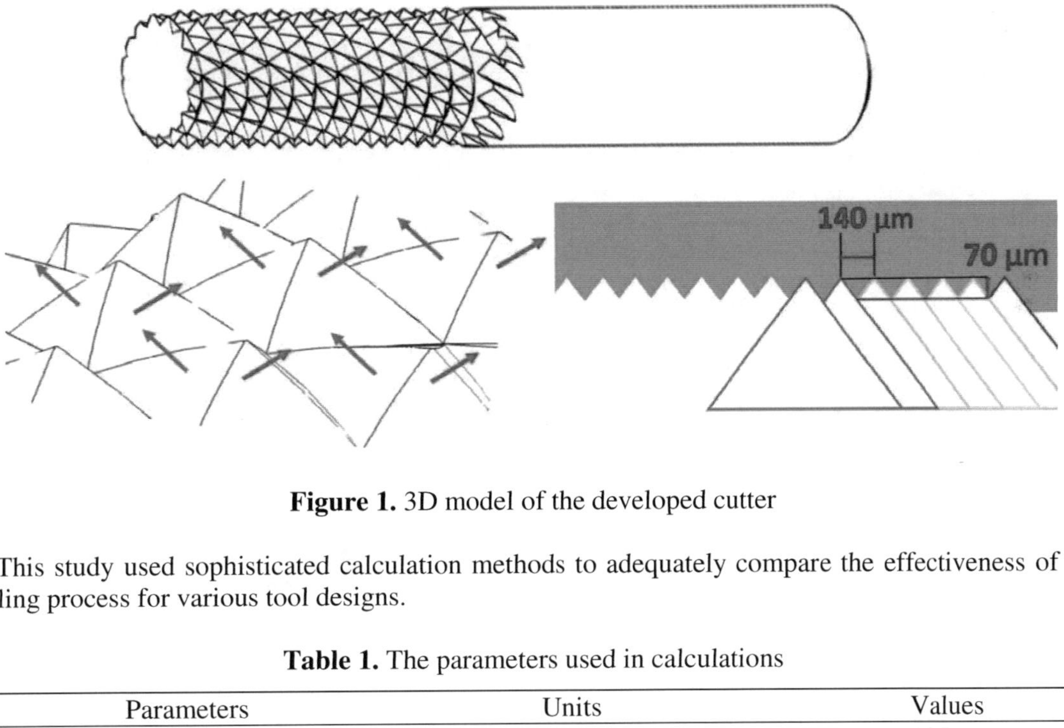

Figure 1. 3D model of the developed cutter

This study used sophisticated calculation methods to adequately compare the effectiveness of the milling process for various tool designs.

Table 1. The parameters used in calculations

#	Parameters	Units	Values
1.	cutting speed	m/min	125
2.	feed per tooth	mm/ tooth	0.15
3.	depth of cut	mm	8
4.	cutting width	mm	9.5
5.	tool rotary speed	rpm	4000

Analytical calculation formulae for determining the volume of cut material per minute (Q, cm^3/min) and the number of cutting cycles per scond (N, s^{-1}) used (formulae 1 - 3).

$$Q = a_p \times a_e \times V_f / 1000 \tag{1}$$

a_p – depth of cut, mm;
a_e – cutting width, mm;
V_f – feed per minute, mm/min.

$$V_f = n \times f_z \times z_n, \tag{2}$$

n – tool rotary speed, rpm;
f_z – feed per tooth, mm/tooth;
z_n – number of tooth.
The number of cuts per second (N, s^{-1}) in this case can be determined by formula 3.

$$N = n \times z / 60, \tag{3}$$

z – number of cutting edges per tool diameter at height of 9.5 mm;
Parameter z for mills: z_1 for ISCAR = 50 cutter edges; z_2 for SANDVIK = 3 cutter edges;
The range of cutting conditions during the calculations:
v = 125 m/min or approximately n = 4000 rpm (for mill diameter D = 10 mm);
f_n = 0.15 mm/tooth feed per tooth;
a_p = 8 mm and ae = 9.5 mm depth of cut and cutting width respectively.
The ranges of cutting modes were:

- cutting speed 100 – 200 m/min;
- feed per tooth 0.01 – 0.02 mm/tooth.

The calculation data of the efficiency of various mill models according to parameters of Q and N depending on the cutting speed are presented in the diagrams in figure 2.

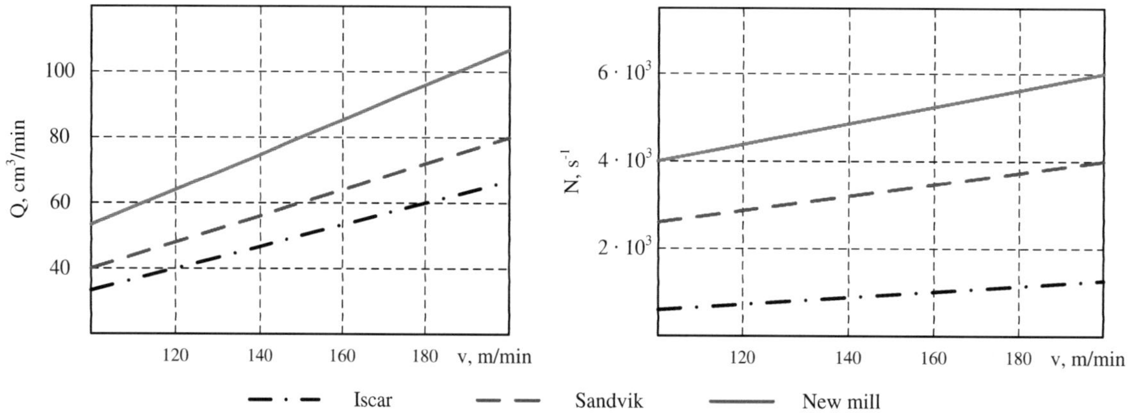

Figure 2. The effect of cutting speed on the volume of the material to be removed and on the number of cuts per second

The calculation data show that the intensity of deformations during the processing using the new geometry significantly increases in contrast to analogical mills. The number of single-chip cutting microcycles varies within the interval: the new mill – 4000-6000 s^{-1}; SANDVIK – 2500-4000 s^{-1}; ISCAR – 800-1500 s^{-1}.

3. Results of simulation and discussion

The conventional analytical calculation models exclude the geometrical features of the new mill geometry; that is why when determining the numerical values of the cutting forces we use the calculation formulae obtained based on the full-scale experiments. However, the nature of the load distribution and the numerical value of the specific cutting forces can be determined using the finite element analysis

IOP Conf. Series: Materials Science and Engineering **754** (2020) 012008 doi:10.1088/1757-899X/754/1/012008

method [7]. A series of numerical experiments using FEM allows determining stresses and deformations in the cut material layer when processing with multi-blade mills of the new type (figure 3), i.e. it allows to indirectly evaluate the unit loads during milling. Besides, the calculation data allow predicting the shape and size of the chips that may appear in case of processing by the new mill.

a) b) c)

Figure 3. The model and the calculation result of the cutting process
(a) tool – detail interface; (b) scheme of cutting and stresses; (c) one cycle of tooth movement and chip formation

Graphic representation of the deformation processes illustrates the kinematics and the sequence of the material layer cutting process. The material is not cut (unlike when standard milling is applied) but it is crushed and removed from the cutting area in the form of microparticles. The new mill works similarly to an abrasive tool combining the processes of cutting and crushing. The studies based on the FEA method show that the specific load when using the new mill geometry is reduced, whereas the maximum values of stresses arising on the surface of a workpiece and the cutting part of the tool do not exceed 500-5000 MPa (figure 3 b, c). In this case, the maximum stress values concentrate in a thin surface interface layer. This fact, as well as the predicted finely crushed type of chips, is a prerequisite for reducing the roughness of the processed surface and at the same time increasing the productivity of milling. The required shape and dimensions of the cutting edge were determined by given various values of the geometric parameters of the cutting part, the properties of the workpiece material and cutting conditions. As a result, 3D model of the end mill was created having the original tooth profile in the form of trapezia (figure 4), located along two intersecting helical lines with tilt angles of 22 and 85 degrees.

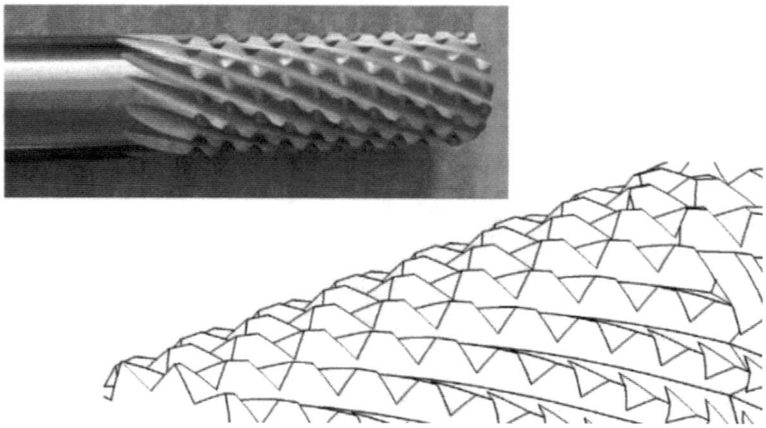

Figure 4. Actual geometry of the cutter tooth's

The manufacturing of tools on ANCA RX7 cutter grinding machine was controlled by a specially developed controlling software. We used rods made of the extruded nanostructured hard-metal composite as a workpiece material for the tool manufacturing. This new hard-metal composite has a number of improved physical and mechanical properties in terms of strength, crack and heat resistance due to the modification of the cobalt binder with the addition of aluminium oxide nanoparticles [8]. A set of these properties is a necessary prerequisite for the effective operation of the developed design of a multi-blade mill at high cutting speeds and under conditions of alternating cyclic loads. The new contour tool was used for testing when cutting composite fibre materials.

4. Verification by experiments

As the main parameter for the experimental verification of the analytical results obtained from the research, we used the surface roughness values of the machined new mill construction in different cutting conditions when processing composite materials. Such heterophase materials cause particular processing difficulties since, as a result of structural heterogeneity on the machined surface, chips and delaminating (destruction of the binder material) appear, as well as fibre fragments elongated from the bulk of the workpiece. For these reasons, fibre composite materials can be relevant objects to test the efficiency of the new mill design. Along with the improved quality of the processed surface, the experimental studies also resulted in finding a change in the chip shape, where it has a finely crushed structure, with the size of the elements (chip fragments) of about 2 μm (figure 5 c), which is in good agreement with the results of the preliminary FEM estimates.

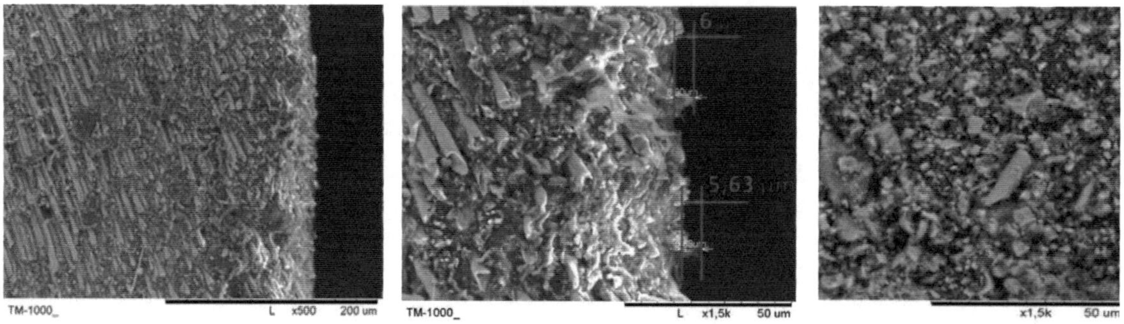

a) b) c)

Figure 5. Electron microscope images of the surface topography (a – b) and chips (c)

The surface quality at processing with original milling cutter is obviously higher than at processing with a standard one. It has been ascertained that the values of surface roughness significantly decrease and these values are lower than 6 μm (figure 5 b). It can be explained by the fact that due to the increase in cutting edges and intensity of cutting processes (the number of single cutting cycles is up to 4000-6000 s^{-1}), the composite fibres are easily cut and are not elongated from the matrix material. The chip morphology and sizes correspond to the cutting intensity specific to high-speed micromilling. The chip crushing coefficient at a milling depth of 0.5 mm and a rotational speed of 4000 rpm is equal to k = 250. This fact indirectly confirms the increased productivity of the machining process and a decrease of the specific cutting force.

5. Conclusions

Using conventional analytic models for calculation of deformation processes and methods of final element analysis (FEA), we studied the influence of the geometry of the cutting part of the multi-edge tool and cutting modes on the intensity of the deformation processes of high-speed milling. On this basis, we determined the rational design of the contour mill, and the experimental studies confirmed the efficiency and increased productivity with simultaneous quality improvement when machining by the new mill.

References

[1] Hideaki O, Koji U, Ippei K, Junichi H and Yasuhiro N 2015 High speed processes with long oblique cutting edges *Journal of Manufacturing Processes* **19** 95–101

[2] Binchurov A S et al. 2019 Influence of cutting modes on power characteristics of rotational turning by multifaceted cutters *IOP Conf Ser Mater Sci Eng* **537** 032101

[3] Brandao de Oliveira F, Rodrigues A R, Coelho R T and Fegali de Souza A 2015 Size effect and minimum chip thickness in micro milling *International Journal of Machine and Manufacture* **89** 39-54

[4] Cuba Ramos A, Autenrieth H, Strauß T, Deuchert M, Hoffmeister J and Schulze V 2012 Characterization of the transition from ploughing to cutting in micro machining and evaluation of the minimum thickness of cut *J. Mater. Process. Technol* **212** 594–600

[5] Vogler M P, Devor R E, Kapoor S G 2004 On the modeling and analysis of ma-chining performance in micro-end milling Part I: Surface generation *J. Manuf. Sci. Eng* **126** 685–694

[6] Chuzhoy L, Devor R E, Kapoor S G, Bammann D J 2002 Microstructure-level modeling of ductile iron machining *J. Manuf. Sci. Eng* **124** 162–169

[7] Mohammad L, Saeid A, Mohsen A 2018 3D FEM simulation of tool wear in ultrasonic assisted rotary turning *Ultrasonics* **88** 106-114

[8] Gordeev Y I et al. 2019 Study of the formation features of hard metal composites structure obtained from bimodal powder mixtures *IOP Conf Ser Mater Sci Eng* **511** 012032

IOP Conf. Series: Materials Science and Engineering **754** (2020) 012009 doi:10.1088/1757-899X/754/1/012009

Hydrochemical Research and Geochemical Classification of Salt Lakes in the Pavlodar Region

A Ubaskin[1], A Kassanova[1,2], A Lunkov[1], K Ahmetov[1], K Almagambetova[1], N Erzhanov[1] and T Abylkhassanov[1]

[1]S. Toraighyrov Pavlodar State University, 64 Lomov Street, Pavlodar, 140000, Kazakhstan
[2]National Research Tomsk Polytechnic University, 30 Lenin Avenue, Tomsk, 634050, Russia

E-mail: asiyakass@mail.ru

Abstract. This paper presents the results of a study of the macrocomponent composition of 15 salt lakes of the Pavlodar region. For the first time, based on the data of the chemical composition and pH of the waters, the geochemical classification of the lakes of this region has been proposed. The study shows that the majority of the analyzed objects are lakes of chloride type with sodium cationic composition.

1. Introduction
Salt lakes are important sources of mineral salts - mirabilite, halite, soda, gypsum and others [1-3], in addition, recreational use of the territory of salt lakes is highly popular [4, 5]. Salt lakes that are unique in their physicochemical properties are rarely lifeless, their biota is represented by specific halophilic forms, in particular, to the crustacean Artemia Leach, 1819 [6-8]. Artemia is widely used as a starter feed for growing fish, crustaceans, farm birds [9, 10] in industrial conditions, for obtaining valuable medicinal and cosmetic products [11].

In the hydrochemical study of natural waters, an important issue is their classification. According to the Kurnakovo-Valyashko classification, salt lakes are usually divided into carbonate, sulfate and chloride chemical types. Sometimes they are also distinguished by their acidic and basic types, sodium and magnesium subtypes. Lakes with a pH of more than 9.0 are usually classified as basic types, lakes with a pH less than 9.0 and the predominant Cl^- anion are typical of chloride types; lakes with a pH less than 9.0 and the predominant SO_4^{2-} anion are typical of sulfate types [12, 13]. Hydrochemical characteristics of salt lakes are subject to significant chronological changes. The reason for the change in hydrochemical characteristics is cyclic fluctuations of climatic conditions, leading to periodic filling and drying of lakes [14,15].

Numerous brackish and salty lakes are widespread in the Pavlodar region. The ecosystems of these reservoirs located in the Irtysh plain and the Kazakh Uplands function in transgressive-regressive cycles [16]. Significant climate and water regime variability over years and seasons causes instability, both of individual components of aquatic ecosystems, and of all of them in total. Therefore, it is very important to study their geochemical typing, both from a scientific point of view, and for the development of recommendations for the protection and rational use of lake minerals and valuable hydrobionts.

Thus, the purpose of this study is to explore the chemical composition of salt lakes in the Pavlodar region, to establish the type of lakes and the seasonal change in the type of lakes over the period spring-autumn 2018.

It is known that the spatial arrangement of lakes of different salinity and chemical types does not have a definite pattern. Lakes with different water composition and salinity can often be located at a small distance from each other. In this regard, when choosing the lakes for hydrochemical analysis, we were guided, firstly, by the value of the lake for the commercial use of brine shrimp resources, and secondly, by lakes, with abiotic factors limiting the existence of brine shrimp populations, and thirdly, by geographical location (to cover most of the district centers) (table 1).

Table 1. The name and location of the investigated salt lakes

Name of the lake	Geographical coordinates	S, km^2	Water Regime	Ice Regime
Bastuz	52°39′ N, 75°01′ E	7.7	Dries out periodically	Does not freeze
Tuz	50°34′ N, 75°13′ E	7.1	Does not dry out	Gets frozen
Big Kosor	51°18′ N, 78°15′ E	2.9	Dries out periodically	Gets frozen
Muzdykol	53°43′ N, 76°02′ E	1.7	Dries out periodically	Does not freeze
Bayanbay	53°00′ N, 74°37′ E	2.8	Does not dry out	Gets frozen
Kalatuz	51°53′ N, 77°30′ E	8.5	Does not dry out	Does not freeze
Borly	51°49′ N, 77°58′ E	16.0	Does not dry out	Does not freeze
Aydarsha	51°45′ N, 78°48′ E	2.8	Does not dry out	Gets frozen
Seiten	51°55′ N, 78°07′ E	15.5	Does not dry out	Does not freeze
Zhamantuz	51°33′ N, 77°44′ E	2.1	Dries out periodically	Does not freeze
Uyaly	51°18′ N, 78°15′ E	3.8	Does not dry out	Gets frozen
Sharbakty	51°23′ N, 78°15′E	6.8	Does not dry out	Does not freeze
Balkazy	53°10′ N, 77°03′ E	4.8	Dries out periodically	Does not freeze
Kyzyltuz	51°52′ N, 75°58′ E	10.7	Does not dry out	Does not freeze
Kudaikul	51°39′ N, 78°11′ E	6.2	Does not dry out	Does not freeze

2. Results and discussions

Sampling and preservation of water was carried out according to the standard technique [17]. The total chemical composition of the water was determined in an accredited laboratory of the Kazakhstan Project-Research Institute "Kazakhstan Project" LLP. The total mineralization index was determined by the gravimetric method [17]. A potentiometric method was used to determine the pH of the samples. The content of Cl^-, NO_3^-, NO_2^-, PO_4^{3-} and SO_4^{2-} ions was determined by capillary electrophoresis [19], the content of Na^+, K^+, Ca^{2+}, Mg^{2+}, NH_4^+ cations by capillary electrophoresis according to [20]. The permanganate oxidizability, the total content of CO_3^{2-} and HCO_3^- ions were performed by titrimetric methods of analysis [18]. Determination of total iron content was performed by photocolorimetric method using sulfosalicylic acid [18]. For the analysis were used reagents brand "pure for analysis".

The results of the chemical analysis of the composition of water samples of salt lakes of Pavlodar region in the spring-autumn 2018 period are presented in table 2.

Table 2. The results of the chemical analysis of the composition of water samples of salt lakes

Name of the lake	Season	pH	Total minera-lisation	Fe	PO	Cl^-	SO_4^{2-}	HCO_3^-	CO_3^{2-}	PO_4^{3-}	Na^+	Ca^{2+}	Mg^{2+}	K^+
			g/l	mg/l		g/l				mg/l	g/l			mg/l
Bastuz	spring	7.70	190.5	0.06	29.7	104.3	19.10	0.37	<0.008	<0.25	59.59	0.670	6.39	<0.5
	autumn	7.61	275.8	0.11	25.5	142.1	26.53	0.50	<0.008	<0.25	96.53	0.0004	10.14	<0.5
Tuz	spring	8.21	39.1	0.07	19.7	22.0	4.54	0.36	0.096	<0.25	10.42	0.083	1.32	<0.5
	autumn	8.60	110.5	0.07	28.2	12.0	3.24	0.34	0.072	<0.25	93.40	0.190	1.15	46.2
Big Kosor	spring	7.72	73.3	0.09	23.8	40.6	4.51	0.23	<0.008	<0.25	19.41	1.800	5.96	<0.5
	autumn	7.50	199.8	0.12	28.2	122.6	8.42	0.38	<0.008	<0.25	51.62	0.770	15.98	<0.5
Muzdykol	spring	7.72	200.1	0.10	31.6	104.5	19.10	0.39	<0.008	<0.25	61.54	0.730	13.18	<0.5
	autumn	7.32	319.2	0.10	32.6	188.1	29.44	0.51	<0.008	<0.25	84.28	0.0004	16.81	<0.5

Bayanbay	spring	7.47	292.2	0.12	23.0	14.3	4.32	0.35	<0.008	<0.25	8.65	0.260	1.05	28.5
	autumn	7.30	356.7	0.10	30.3	176.5	53.06	1.15	<0.008	<0.25	105.50	0.0004	20.49	<0.5
Kalatuz	spring	7.92	167.4	0.19	28.3	90.1	7.98	0.49	0.1350	<0.25	55.24	0.410	11.99	<0.5
	autumn	7.40	192.1	0.09	31.0	104.4	11.35	0.61	<0.008	<0.25	62.80	0.300	12.67	<0.5
Borly	spring	9.16	57.2	0.08	24.8	26.7	5.49	2.32	1.380	<0.25	21.19	0.009	0.17	<0.5
	autumn	8.89	62.6	0.06	26.7	26.0	6.22	2.81	1.200	<0.25	26.17	0.012	0.20	<0.5
Aydarsha	spring	8.50	68.9	0.05	6.1	22.9	5.88	0.66	<0.008	<0.25	70.71	0.084	0.98	494
	autumn	8.70	74.4	0.04	6.4	36.9	7.81	1.01	<0.008	<0.25	64.11	0.057	0.92	442
Seiten	spring	7.80	115.9	0.05	3.7	49.4	6.47	1.04	<0.008	<0.25	99.73	0.045	0.13	365
	autumn	8.00	124.8	0.04	5.4	59.5	11.40	1.22	<0.008	<0.25	100.30	0.019	0.35	422
Zhamantuz	spring	9.00	157.4	0.05	4.6	67.8	17.11	0.22	<0.008	<0.25	101.40	0.430	2.07	479
	autumn	9.10	215.6	0.04	5.7	101.6	22.70	0.36	<0.008	<0.25	132.90	0.570	2.01	1012
Uyaly	spring	9.50	23.4	0.09	24.9	2.7	3.06	3.59	0.910	<0.25	11.99	0.015	0.08	17
	autumn	9.47	27.3	0.07	23.8	4.8	6.49	2.75	1.500	<0.25	11.62	0.0056	0.09	48
Sharbakty	spring	8.95	57.4	0.10	28.2	12.9	0.82	1.49	0.420	<0.25	40.56	0.020	0.43	11
	autumn	8.77	119.0	0.06	28.0	67.1	5.92	1.39	0.750	<0.25	43.47	0.004	0.54	<0.5
Balkazy	spring	8.40	131.1	0.10	25.0	32.1	51.28	0.51	0.096	<0.25	42.03	0.470	3.76	<0.5
	autumn	8.35	145.5	0.06	29.2	48.0	41.76	0.74	0.048	<0.25	47.77	0.296	6.92	<0.5
Kudaikul	spring	7.39	246.2	0.13	27.3	141.4	20.27	0.31	<0.008	<0.25	73.73	0.600	9.84	<0.5
	autumn	7.12	368.1	0.12	31.2	204.1	57.77	0.29	<0.008	<0.25	122.20	0.0004	15.49	<0.5
Kyzyltuz	spring	7.50	239.9	0.90	6.6	119.7	17.16	0.20	0.120	0.12	145.90	0.258	2.29	652
	autumn	9.00	282.6	0.04	5.7	252.7	22.39	0.42	<0.008	0.14	150.00	0.211	2.00	848

The content of total iron in the analyzed objects varies from 0.04 to 0.19 mg/l and varies slightly from spring to autumn and meets the standards for the maximum permissible concentration for water of reservoirs of the 1-2 categories (maximum permissible concentration of maximum 0.3 mg/l). In most water bodies, the content of nitrate and nitrite ions does not exceed 0.2 mg/l and does not change within six months, with the exception of the lakes Aydarsha and Zhamantuz, in which by the autumn the content of nitrate ions increases 13 and 57 times respectively, while according to the standards for water of water bodies of categories 1 and 2, their content does not exceed the norm (MPC_{NO2} not more than 3.3 mg/l, MPC_{NO3} not more than 45.0 mg/l). The content of phosphate ions in all salt lakes is less than 0.25 mg/l and does not vary during the spring and autumn periods.

Indicator content of organic substances is permanganate oxidation (PO). The minimum value of PO is observed for Lake Seiten and is 3.7 mg/l (spring), the maximum value is 32.6 mg/l and is typical for Lake Muzdykol (autumn). As a rule, the permanganate oxidation slightly rises from spring to autumn and is directly dependent on the total salinity. For lakes with high total mineralization, PO has the greatest value, lakes with lower mineralization have lower PO.

For further geochemical classification of salt lakes in the Pavlodar region, an important role is played by such indicators as pH, total mineralization, the content of anions Cl^-, SO_4^{2-}, CO_3^{2-}, HCO_3^-; Na^+, Ca^{2+}, Mg^{2+} cations. It is these cations and anions that are the basis of the macro-component analysis of lakes.

The total mineralization of water in lakes varies from 23 to 268 g/l in the spring, and significantly increases in the autumn from 27 to 367 g/l (figure 1).

According to the degree of mineralization, the lakes are divided as follows:

- slightly salted (Lake Uyala (spring, autumn), Lake Bayanbai (spring));
- highly salty (Lake Tuz (spring));
- slightly concentrated brines (Big Kosor (spring), Borley, Aydarsha, Sharbakty (spring));
- strong brines (Kalatuz, Seyten, Zhamantuz, Balkazy, Kyzyltuz);
- very strong brines (Bastuz and Muzdykol during the autumn period);
- super strong brines (Kudaykul and Bayanbay in autumn).

The findings show that the optimal levels of mineralization for the growth of the crustacean population are in the range from 34 to 299 g/l, in rare cases 320 g/l [20].

The pH value in the spring period ranges from 7.39–9.5. During the autumn period in most of the analyzed objects, the pH level rises or decreases slightly (figure 2).

The highest pH values are characteristic for lakes with lower mineralization, for example, the pH of Borly and Uyaly lakes are more than 9, and the total salinity is less than 60 g/l. The minimum pH values (7.12-7.3) are fixed in brine-type lakes. In all the lakes studied, this trend continues, with the exception to this rule being Lake Zhamantuz, which has a pH of more than 9 with a high level of salinity (more than 250 g/l).

It is known that the indicator of mineralization of natural waters is directly dependent on the content of such anions as Cl^-, SO_4^{2-}, CO_3^{2-} and HCO_3^-, and the pH of natural waters depends on the concentration of carbonate and bicarbonate ions [21].

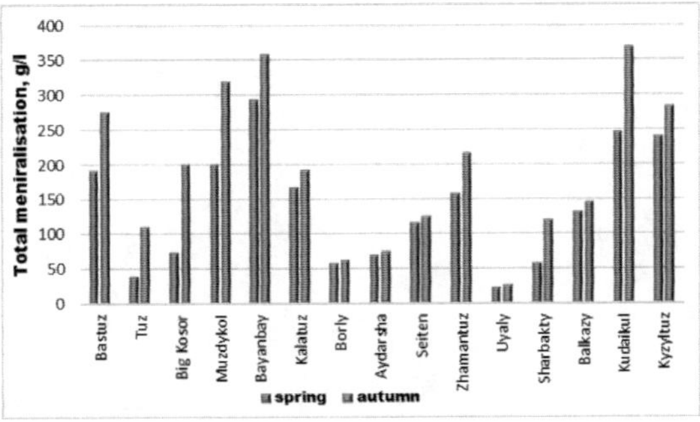

Figure 1. Graph of changes in the total mineralization of lakes in the period spring - autumn 2018.

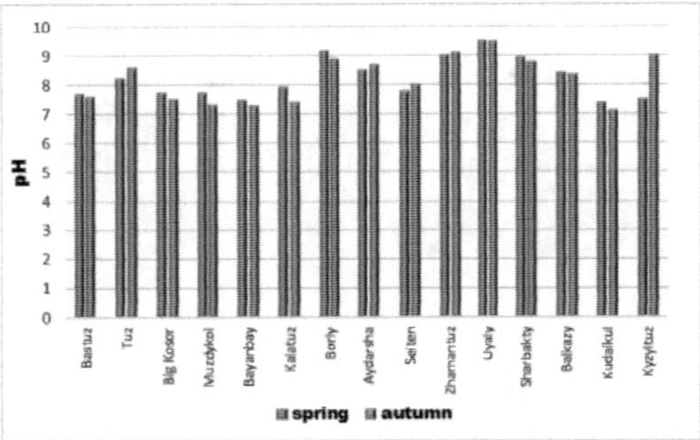

Figure 2. Graph of changes in the pH of the analyzed objects for the period spring-autumn 2018.

Figure 3 shows a graph of the dependence of the total salinity of lake waters on the content of the above-mentioned ions in the spring.

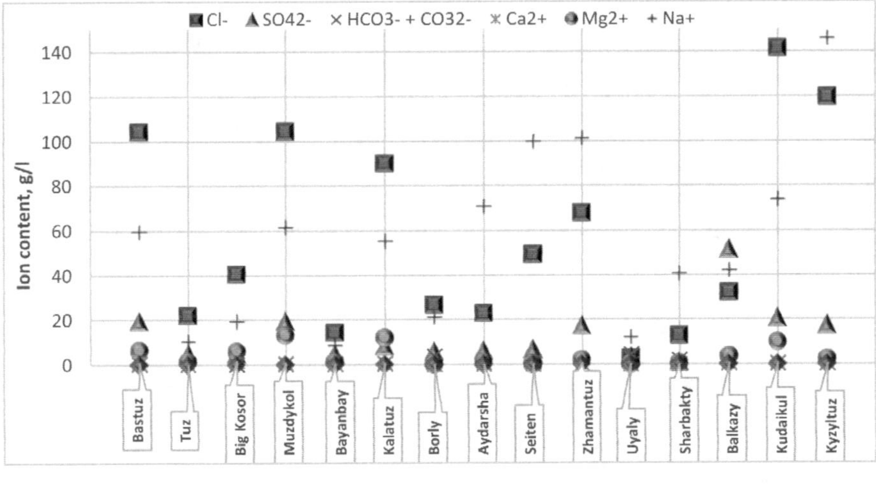

Figure 3. Graph of the content dependence of basic ions on the level of total mineralization in the spring period

As can be seen from the graphs presented, the predominant anion in salt lakes in the Pavlodar region in the spring is Cl^-, so most lakes can be attributed to chloride-type lakes. In spring, two lakes can be distinguished, where the dominant anion is not Cl^-. This is Lake Uyaly, where carbonate and bicarbonate ions are the predominant anion, and Lake Balcas, where the main anion is SO_4^{2-}. At the same time, taking into account the classification proposed in [13], lakes with pH> 9 can be attributed to basic or carbonate lakes. These are the lakes Uyaly and Borley.

To determine the cation composition of the analyzed objects, the dependence of the total mineralization index on the concentration of the main cations: Na^+, Ca^{2+}, Mg^{2+} was built.

According to the classification of lake types, all the studied lakes during the spring period can be divided into the following: carbonate or soda type (with a pH greater than 9) of the sodium subtype - 2 lakes (13% of the total); sulfate type of sodium subtype - 1 lake (7%); chloride type of sodium subtype - 12 lakes (80%).

The directional transformation of the chemical composition of lake waters is due to the concentration in the summer and autumn periods, changes in the chemical composition leads to a change in the types of lakes: from carbonate to sulfate and then to chloride [22]. This sequence of change of the chemical type of lakes is explained by the different solubility and, accordingly, precipitability of salts. First of all, the least soluble calcium and magnesium carbonates precipitate, then calcium, magnesium, sodium sulfates, and finally of all precipitate chlorides from the supersaturated brine.

Given that the Pavlodar region is located in a sharply continental climate zone, we should expect a change in the type of lakes from carbonate to sulfate, from sulfate to chloride. To confirm these hypotheses, a graph was constructed of the dependence of the content of basic ions on the level of total mineralization in the autumn period (Figure 4).

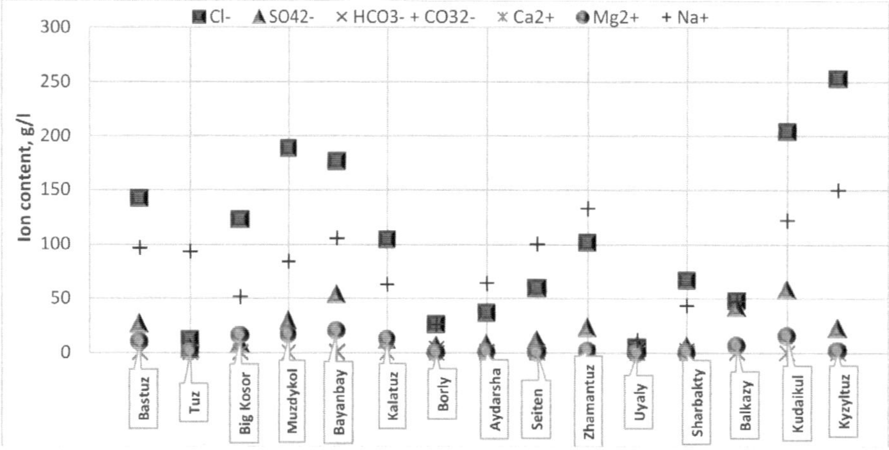

Figure 4. Graph of the content dependence of basic ions on the level of total mineralization in the autumn period

As can be seen from the presented graph, in 14 of 15 analyzed lakes the main anion is the Cl^- ion, one lake (Lake Uyala) has almost the same content of anions, the main anion is SO_4^{2-}, but the pH of the lake is more than 9 and by classification [14] it is carbonate.

The predominant cation, as in the spring, is the sodium cation.

During the autumn period, the following types of lakes can be distinguished: carbonate type of sodium subtype - 1 lake (7% of the total); - sodium chloride-type subtype - 14 lakes (93%).

The results of the geochemical classification of the analyzed salt lakes of the Pavlodar region for the purpose of clarity were summarized in Table 3.

Table 3. Geochemical types of salt lakes of Pavlodar region in the period spring-autumn 2018

Name of the lake	Geo-chemical type	
	Spring	Autumn
Bastuz	Chloride-sodium Cl-Na-II	Chloride-sodium Cl-Na-II
Tuz	Chloride-sodium Cl-Na-II	Chloride-sodium Cl-Na-II
Big Kosor	Chloride-sodium Cl-Na-III	Chloride-sodium Cl-Na-III
Muzdykol	Chloride-sodium Cl-Na-II	Chloride-sodium Cl-Na-II
Bayanbay	Chloride-sodium Cl-Na-II	Chloride-sodium Cl-Na-II
Kalatuz	Chloride-sodium Cl-Na-I	Chloride-sodium Cl-Na-II
Borly	Carbonate-sodium C-Na-I	Chloride-sodium Cl-Na-I
Aydarsha	Chloride-sodium Cl-Na-II	Chloride-sodium Cl-Na-I
Seiten	Chloride-sodium Cl-Na-I	Chloride-sodium Cl-Na-I
Zhamantuz	Chloride-sodium Cl-Na-II	Chloride-sodium Cl-Na-II
Uyaly	Carbonate-sodium C-Na-I	Carbonate-sodium C-Na-I
Sharbakty	Chloride-sodium Cl-Na-I	Chloride-sodium Cl-Na-I
Balkazy	Sulfate-sodium S-Na-II	Chloride-sodium Cl-Na-II
Kudaikul	Chloride-sodium Cl-Na-II	Chloride-sodium Cl-Na-II
Kyzyltuz	Chloride-sodium Cl-Na-I	Chloride-sodium Cl-Na-I

Thus, the study of the hydrochemical composition and geochemical classification of salt lakes in the Pavlodar region was conducted for the first time. Most of the objects analyzed belong to the lakes of the sodium-type sodium subtype. On the territory of the region there are several carbonate and sulphate type lakes, characterized by a gradual transition to chloride type lakes.

The results of the research will make it possible to use these lakes extensively both for the extraction of sodium mineral salts and for the development of Artemia crustaceans. In addition, data on the geochemical classification of saline lakes will allow to predict the nature of groundwater feeding the lakes; their geology, formation and other worth-exploring issues.

References

[1] Sklyarov E V, Sklyarova O A, Men'shagin Yu V and Danilova M A 2005 *Geography and Natural Resources* **32** 323–332
[2] Sovolov V P Bodies of Altai Territory: Biological Productivity and Prospects for Use 1999 *Novosibirsk Nauka* 285
[3] Novoselova S I and Novoselov V A 2000 *Sib. Ecolog. J.* **3** 80–85
[4] Pasynkov A A, Sockova L M and Chaban V I 2014 *Scientific notes of the Tauric National University named after V I Vernadsky Geography* **27** 97–117
[5] Vasil`ev O F, Kazancev V A, Popov P A and Kirillov V V 2005 *Sib. Ecolog. J.* **2** 167–173
[6] Ivanova V I 2012 *X International Scientific and Practical Conference, ICSN "Science and Education* 38–40
[7] Volf L A 2011 Ecological features of the gill-footed crustacean (Artemia parthenogenetica Varigozzi, 1974) in the salt reservoirs of Northern Kazakhstan (Republic of Kazakhstan) *Omsk* 17
[8] Ubaskin A V 2005 Ecological and biological features of the crustacean Artemia (Crustacea, Anostraca) of salt lakes in Pavlodar region (Republic of Kazakhstan *Omsk* 21
[9] Litvinenko L I, Kozlov A V, Matveeva E P and Guzhenko M V 2006 *Bulletin of KSU Natural Sciences* **4** 49– 51
[10] Studenikina T L 1984 Biological resources of inland waters of Siberia and the Far East *Moscow Nauka* 117–124.
[11] Litvinenko L I 2009 Gill-footed crustaceans of the genus Artemia Leach, 1819 in the hyperhaline reservoirs of Western Siberia (geography, biodiversity, ecology, biology, and practical use) *Perm`* 46
[12] Shirokova V A 2013 *Bulletin of TSU* **18** 1023–27
[13] Kolpakova M N 2014 Geochemistry of salt lakes in Western Siberia *Tomsk* 21
[14] Shwarcev S I, Kolpakova M N, Isupov V P, Vladimirov A G and Ariunbileg S 2014 *Geochemistry* **3** 1-18
[15] Kolpakova M N, Borzenko S V and Isupov V P 2015 *Water: chemistry and ecology* **1** 18–23
[16] Ubaskib A V and Ubaskina N V 2002 *History, nature, economics. Mater. Int. scientific-practical conf.(Omsk)* 188–191
[17] Leonova G A, Bobrov V A, Bogush A A, Bychinsky V A and Anoshin G N 2007 *Geochemistry* **10** 1114–28
[18] GOST 31861-2012. Water. General sampling requirements
[19] GOST 26449.1-85. Stationary distillation desalination plants. Methods of chemical analysis of salt water
[20] KZ.07.00.01998-2014. MVI of the mass concentration of chloride ions, nitrite ions, sulfate ions, nitrate ions, fluoride ions and phosphate ions in samples of natural, drinking and treatment waste water using the capillary electrophoresis system Kapel
[21] KZ.07.00.01529-2017. MVI of the mass concentration of cations of ammonium, potassium, sodium, lithium, magnesium, strontium, barium, calcium in samples of drinking, natural (including mineral) and wastewater using the capillary electrophoresis system Kapel
[22] Litvinenko L I and Guzhenko M V 2007 *Siberian Bulletin of Agricultural Science* **2** 81–85

IOP Conf. Series: Materials Science and Engineering **754** (2020) 012010 doi:10.1088/1757-899X/754/1/012010

Experimental analysis of the EDM process of dissimilar materials

T Ablyaz[1], E Shlykov[1], K Muratov[1], S Oglezneva[1], V Karmanov[1], I Osinnikov[1], S S Sarabjeet[2] and D Sahil[3]

[1] Perm National Research Polytechnic University, Perm-614000, Russian Federation
[2]Mechanical Engineering Department, Beant College of Engineering & Technology, Gurdaspur-143521, India
[3]Thapar Institute of Engineering and Technology, Patiala- 147001, India

[*]E-mail: lowrider11-13-11@mail.ru

Abstract. The aim of the present study is to investigate the surface roughness of Electrical Discharged Machined (EDM) samples with Electrode Tool (ET) materials. For the experiments, three different types of ETs were considered which are made of different materials viz. composite, copper, and graphite. Aluminium, low alloy 35GS steel and OT4 titanium alloy were selected as the workpiece materials. Machining parameters used were: current (I) 8 A, voltage (U) 100 V, and pulse duration (Ton) as 150 μs. It was found that the ET made of composite material performs best in EDM of 35GS steel and OT4 titanium alloy. Specifically it was concluded that in the first case, the performance of composite ET is higher than graphite ET by 59 %. Similarly, in the second case, its productivity was found to be 74% and 95% higher when machined with graphite and copper ET respectively. Moreover, the surface roughness values of 35GS steel machined with composite ET was found to be 15% higher when machined with copper and graphite ET. While processing titanium alloy with composite ET, surface roughness values were observed as high as 44% and 25% when machined with copper and graphite ET respectively.

1. Introduction.

Metals and their alloys with different physical and mechanical properties are widely used in manufacturing a wide range of components. To enhance the properties of components manufactured for aviation, and space technologies, continuous improvement is required in design, materials, and the manufacturing techniques. With the advancement of the technology, the shapes of the components become more complicated to manufacture with the convention al techniques. Materials like titanium and its alloys are widely used in avionic industry due to the properties high strength, light weight, and corrosion resistant to the extreme environments. In such domains, mass and reliability of the material play a vital role. However, titanium alloys are difficult to machine effectively due to the generation of high temperatures in the cutting zones. This high temperature affects the machining efficiency, increase wear rate of the tool and enhance the risk of thermal hardening of the surface of workpiece. Thus, the cutting tools for machining complex geometries of such materials has to be decided on the amount of cutting force being generated. This high bearing capacity of thrust force is difficult to achieve by the conventional machining tools and processes [1, 2].

The solution of this problem can be hybrid machining e.g., electro-physical machining. One of such method is electrical discharge machining (EDM). In today's manufacturing practice, electric discharge machining (EDM) has become the famous, non-traditional, material removal process. That's because of a variety of reasons. EDM allows the machining of extremely hard materials and it is possible to produce complex shapes with high precision. Another feature that meets the expectations of modern manufacturing is its inherent automation capability [3-6]. In EDM, copper and graphite are widely as ET material. Due to the high wear and low productivity achieved by these materials makes EDM process impractical to use for batch manufacturing of aerospace components. A number of wear resistant composite materials have been developed of properties similar to the traditionally used ET materials for machining such components. One of the promising materials for the ET is a composite material: pseudo-alloy (copper-colloidal graphite systems with graphite content of less than 2%).

The influence of the ET material on the surface roughness of the machined surfaces and performance of EDM has not been investigated. Also, Composite material as ET has not been studied completely. Thus, the aim of the study is to investigate the achieved surface roughness and the performance of the EDM of materials with different physical and mechanical properties by using different ETs.

2. Materials and methods

Experiments were conducted on EDM machine, Electronica Smart CNC. The materials selected are mentioned in Table 1. For the experiment, electrodes tool (ET) used were made of different materials: Graphite; Composite Material (Pseudo alloy of the copper + colloidal graphite (CuC) system); Copper (M1 grade). Transformer oil (GOST 982-80) was used as the working fluid. Experiments were performed at a current (I) = 8A, voltage (U) = 100V, and pulse duration Ton = 150μs. Measurements of machining productivity Q(mm/h) were carried out on a coordinate measuring machine (make :Carl Zeiss Contura G2). Surface roughness parameter Ra (μm) was obtained using Mahr Perthometer S2 profilometer as per GOST 2789-73. Surface images were obtained with a microscope (make: Olympus GX51) at a magnification of 100 ×.

Table 1. Different types of workpiece material used in this study.

S.No.	Workpiece material	Specification	Standard
1	Aluminium	AD1	GOST 4784-97
2	Low alloy steel	35 GS	GOST 5781 - 82
3	Titanium alloy	OT4	GOST 19807 - 91

3. Results

The experimental results are presented in Figure 1. Surface roughness achieved by the composite ET is higher than copper and graphite by 59%. While productivity achieved by composite ET was 74% higher than that of a graphite ET, and 95% higher than that of copper ET. However, surface roughness values of 35GS steel machined with composite ET was found to be 15% higher when machined with copper and graphite ET respectively. While machining of titanium alloy with composite ET resulted in low surface roughness values as compared to when machined with copper and graphite ET. When processing aluminum AD1, the composite electrode showed the lowest productivity, as well as the worst roughness of the treated surface. Graphite ET showed the lowest roughness values during the processing of aluminum AD1 and steel 35GS. In addition, when processing aluminum, the graphite electrode exhibits the greatest productivity. Copper ET showed the least surface roughness values for OT4 alloy, but at the same time its productivity is 5 times lower than the graphite electrode and 20 times lower than the performance of composite ET (Figure 2).

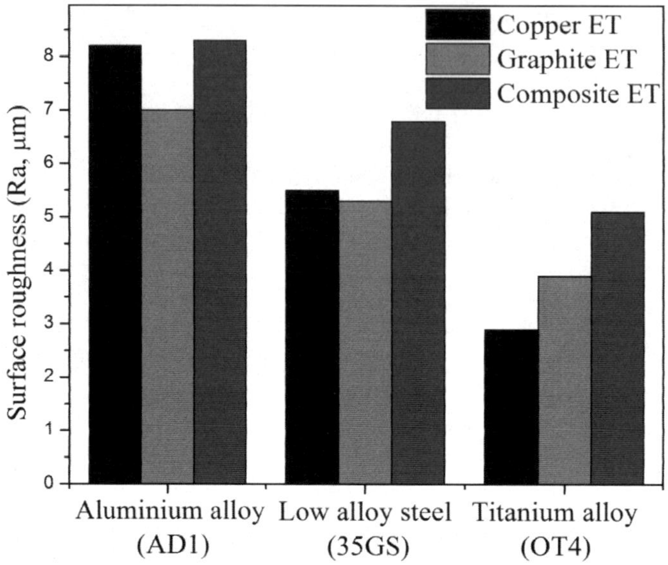

Figure 1. Dependency of the surface roughness parameter (Ra) on ET material.

The study of the surface topography after EDM showed that the surface acquires a specific roughness due to the formation of micro-wells on the surface. Processing with a copper electrode results in a surface with a distinct shape of the micro-wells (Figure 3 (a)). When machined with graphite and composite ETs, the micro-wells combine with each other and form large depressions, which lead to an increase in surface roughness (Figure 3 (a), (b)).

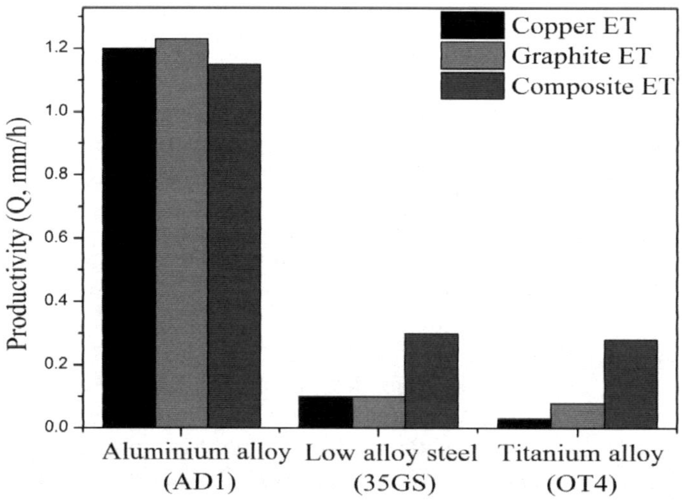

Figure 2. Relationship between productivity (Q) on ET material.

IOP Conf. Series: Materials Science and Engineering **754** (2020) 012010 doi:10.1088/1757-899X/754/1/012010

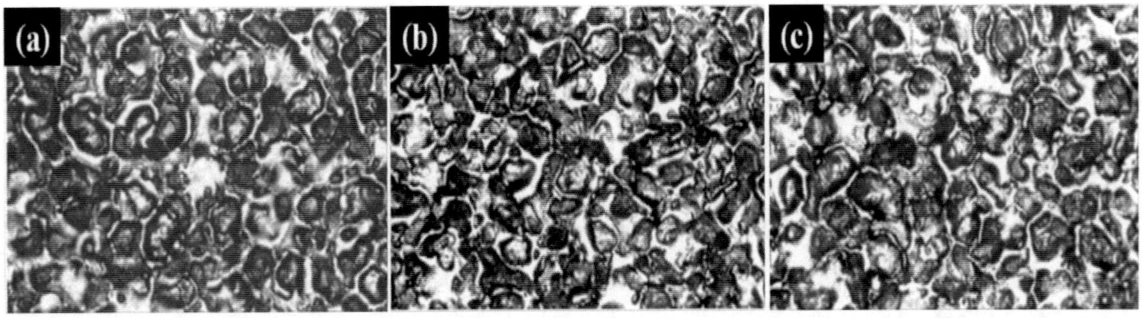

Figure 3. Surface topography of steel 35GS after processing with three types of electrodes, (a) copper ET; (b) graphite ET; (c) composite ET; 100 ×

4. Conclusions

Based on the results of the experiment and analysis of the data obtained, we can draw the following conclusions:

1. The electrodes made of composite material of copper-colloidal graphite system is advantageous to use for EDM of titanium alloys and low alloy steels to achieve very high surface finish.
2. Graphite ET is suitable for the moderate finishing of low alloy steels, titanium alloys, and aluminum.
3. It is advisable to use copper ET to achieve better finishing in EDM of steels.

References

[1] Kovalenko V S 2000 Unconventional methods of processing materials in Japan *Electronic processing of materials* **3** pp. 4-12

[2] Sidhu S S and Bains, P S 2017 Study of the Recast Layer of Particulate Reinforced Metal Matrix Composites machined by EDM. *Materials Today: Proceedings* **4 2** 3243-51

[3] Ablyaz T R, Khanov A M and Khurmatullin O G 2012 *Modern approaches to the technology of electric discharge machining of materials* (Perm: Publishing house Perm. Nat. researched Polytechnic University) 112 p.

[4] Yeliseyev Y and Savushkin B 2010 *EDM of aerospace products* (Moscow: MSTU him NE.)

[5] Smolentsev V P 1983 Electrophysical and electrochemical methods of processing materials: in 2 tons (Higher School) **1** 247

[6] Eliseev Yu S, Savushkin B P 2010 Electroerosive processing of aerospace equipment (M.: Publishing House of MSTU named after N.E. Bauman)

IOP Conf. Series: Materials Science and Engineering **754** (2020) 012011 doi:10.1088/1757-899X/754/1/012011

Density and microstructural investigation of Ce:YAG ceramic subjected to powerful ultrasonic treatment during the compaction process

V D Paygin[1*], E S Dvilis [1], D T Valiev[1], O L Khasanov[1], S A Stepanov[1], V A Vaganov[1], T R Alishin[1], M P Kalashnikov[1, 2]

[1]National Research Tomsk Polytechnic University, Lenin Avenue 30, 634034 Tomsk, Russia
[2]Institute of Strength Physics and Materials Science of Siberian Branch of Russian Academy of Sciences (ISPMS SB RAS), 634055, Russia, Tomsk, pr. Akademicheskii, 2/4

*E-mail: vpaygin@mail.ru

Abstract. The aim of the present work was to investigate the effect of the applied pressure, sintering temperature and ultrasonic treatment during the pressing process on the density and microstructure of translucent ceramics based on yttrium-aluminum garnet doped with cerium ions (Ce: YAG) obtained by conventional sintering of pressed compacts. The optimization of manufacturing conditions of the ceramics was carried out. It was shown that the ultrasonic treatment of initial powder in optimal sintering conditions leads to an increase in the relative density and grain size and decrease in the pore size of the sintered ceramics.

1. Introduction

Ceramics based on yttrium-aluminum garnet - $Y_3Al_5O_{12}$ (YAG) is a promising material owning to it is high melting point and heat resistance, the absence of polymorphic transformations and good electrophysical properties. YAG ceramics can be used as an alternative to glasses or single crystals in devices that work at low visibility, high temperatures, aggressive environments and so one. In past decade, inorganic phosphors based on yttrium-aluminum garnet doped with trivalent cerium ions (Ce:YAG) began to be used for white light-emitting diodes (LED) [1, 2]. However, Ce:YAG-phosphors have amount of disadvantages associated with the degradation of optical and luminescence properties in the process of their use [3]. Ceramic is devoid these disadvantages. Therefore, it is highly relevant to use Ce:YAG-ceramics in white LED-technology [2, 4-10].

The sintering process of high-density ceramics is associated with the problem of keeping the initial phase composition, structure and purity of the material during the pressing process. It is due to the use of various surfactants or binder additives to achieve a higher density of the pressed material without destruction. This is a technological problem that especially relevant for materials used in optical instrumentation where the presence of impurities have a negative impact on the complex of optical-luminescent characteristics. Current problem of high-quality consolidation of Ce:YAG-ceramics can be solved by the methods of hot pressing or spark plasma sintering, however, this leads to high economic costs [2, 11]. It is relevant to use methods that do not require large economic costs and allow obtaining ceramics with uniform density and minimal residual porosity. Such as ultrasonic and collector pressure followed by sintering [12].

In the present work, the possibility of obtaining high-density luminescent Ce: YAG ceramics by ultrasonic pressing followed by sintering in air was demonstrated, the effect of pressure, sintering temperature, and ultrasonic action on the density and microstructure of a consolidated material were investigated.

2. Experiments

Ceramic samples were sintered from a homogeneous mechanical mixture of chemically pure reagents Al_2O_3 (99.99%), Y_2O_3 (99.99%), CeO_2 (99.99%). The powder mixture was made at Chongqing University of Arts and Science (Chonqing, China).

Green bodies were prepared as by the static uniaxial pressing so by the uniaxial ultrasonic pressing [12] at pressure 100-600 MPa on automatic press IP-500AVTO (ZIPO, Russia).

The green bodies were sintered at temperatures from 1550 to 1700 °C for 8 hours in air atmosphere in muffle furnace LHT 02/18 (Nabertherm, Germany).

The sintered ceramic samples had a thickness of 1.5 – 1.7 mm and 8.5 – 9 mm in diameter. Before being characterized, the samples were polished used by polishing machine EcoMet 300 Pro (Buehler, Germany) with Polylab PRO (Kemika, Russia) diamond suspension. The density of the sintered specimens was determined by measuring their dimensions by a digital micrometer “Electron” (Russia) with an accuracy of 10 μm and measuring their mass by a digital balance VLTE-150 (Russia) with an accuracy of 10 mg. In order to gain statistically reliable data, 5 samples were obtained by each processing routes.

The microstructure of the sintered samples were observed by using scanning electron microscope Leo Evo 50 (Carl Zeiss, Germany) with probe microanalysis system “INCA Energy – 350” (Oxford Instruments, England).

YAG-ceramic samples were studied by X-ray diffraction (XRD) phase analysis. The XRD patterns were recorded using XRD-7000S (Shimadzu, Japan).

3. Results and discussion

Figure 1 presents the obtained dependence of the density of green bodies (black curve) and sintered at 1650°C ceramics (grey curve) on the compaction pressure.

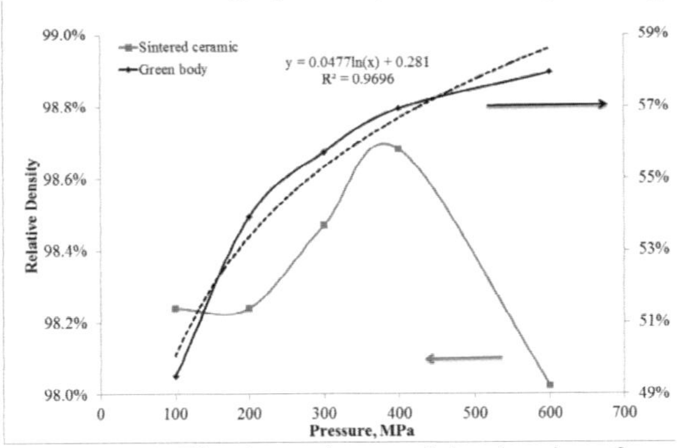

Figure 1. The dependence of the density of Ce:YAG on pressure

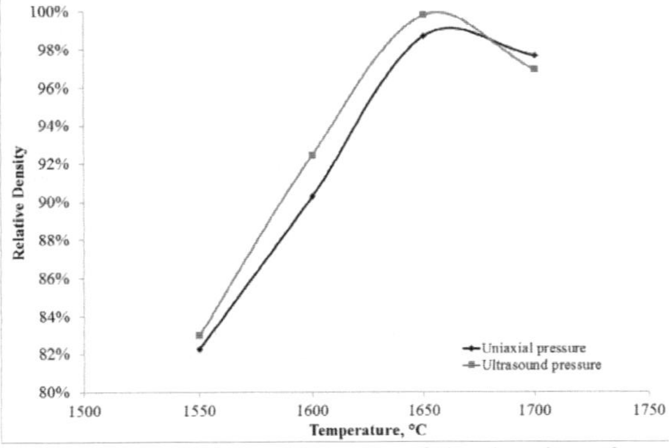

Figure 2. The dependence of the density of Ce:YAG on sintering temperature

The density of the green body with increasing pressure increases according to the logarithmic law [13, 14]. The behavior of the studied material during the densification process in a closed rigid mold with sufficient reliability (~ 97%) can be described by a logarithmic equation in a dimensionless form (1).

$$\rho = b \cdot \ln\left(\frac{P_{np}}{P_{\kappa p}}\right) + 1, \tag{1}$$

where ρ – relative density of the green body, b – is a material constant defining densification pressure sensitivity, P_{pr} is the applied compaction pressure, P_{cr} is the critical pressure (pressure enabling compaction up to full density of the powder material). The coefficients a logarithmic equation were determined by approximating the experimental compaction curves with approximation certainty not less than 96%, the value of coefficient b for the investigated material was 0.0477.

An increase in the compaction pressure from 100 to 600 MPa leads to an increase in the density of the green body from 49.51% to 57.94%. The maximum density value (98.68%) after sintering (figure 1) is observed in ceramics obtained at compaction pressure – 400 MPa. In sintered ceramic samples obtained at compaction pressures above 400 MPa are formed the cracks that are visible to the naked eye. And in some green bodies, obtained at similar pressures lamination was observed. Apparently, this is caused by anisotropic density distribution over the volume during the compaction process of green body which leads to the accumulation of residual stresses and formation of macrodefects.

Based on the obtained results, further molding of the samples was carried out under the pressure of 400 MPa.

Figure 2 presents the obtained dependence of the density of sintered ceramics on the sintering temperature.

The maximum density value 98.7 % and 99.8% is observed for samples obtained by uniaxial and ultrasonic pressing with followed sintering at 1650 °C, respectively. An increase in sintering temperature from 1550 °C to 1650 °C leads to leads to significant rise the relative density of samples by 16.4%. Further increase in sintering temperature from 1650 °C to 1700 °C leads to decrease the relative density by 1 - 2.9%.

Table 1. Density and microstructure properties of sintered Ce:YAG-ceramic

Sample	Relative density, %	Average grain size, μm	Average pore size, μm
1550	82.26±1	1.95	0.49
1550*	82.99±1	1.6	0.54
1600	90.26±1	2.19	0.62
1600*	92.44±1	2.2	0.64
1650	98.68±1	4.2	1.25
1650*	99.78±1	4.57	0.94
1700	97.65±1	15.4	1.8
1700*	96.90±1	13.8	1.19

**Ultrasonic pressing*

Figure 3 shows SEM-images of the spall microstructure of ceramics sintered at different temperatures. It was found that the fracture behavior of ceramics is changes from intercrystalline at temperatures 1550-1600 °C transcrystalline at temperatures 1650-1700 °C. This indicates an increase in the strength of grain boundaries to a value comparable to the grains strength. The measurement results of the relative density, average grain and pore size of investigated ceramics represented in the table 1.

An increase in sintering temperature from 1650 °C to 1700 °C leads to an increase average grain size of investigated ceramics from 1.95 to 15.4 μm, average pore size increase from 0.49 to 1.8 μm. This kind of the change in the average size of structural elements can be explained as follows. At sintering temperatures of 1550-1600 °C with an isothermal exposure - 8 hours and relatively low heating rates, the sintering process does not seem to be completed, and therefore the grain growth process proceeds slowly, and their size remains at the level of the agglomerates of initial powder. Density of the samples for this type of ceramics is relatively low and does not exceed 92.5 %. At sintering temperature - 1650 °C with an isothermal exposure - 8 hours, densification of sintered material occurs intensively and accompanied by decreasing pore quantitative content and grain growth. At a temperature of 1700 ° C and an isothermal exposure time - 8 hours, sintering process significantly slows down, which result in a significant grains growth without further densification of material, which leads to the transition of intergranular residual pores into the volume of grains.

The influence of ultrasonic treatment at relatively low sintering temperatures (1550-1600 °C) is weakly expressed. The effect of ultrasonic treatment can be fully evaluated at temperatures 1650 °C and 1700 °C. Thus, the residual pores average size of the samples obtained by ultrasonic pressing at sintering temperature 1650 °C on 0.31 μm less, the average grain size on 0.37 μm less and the density on 1.1% higher than that of the samples obtained by uniaxial pressing followed by conventional sintering at the same temperature. Increasing density and grain size of the sonicated ceramic samples can be explained by activation effects of ultrasonic treatment, whereby the grain growth speed is enhanced. Sonicated ceramic samples demonstrate the predominantly transcrystalline fracture behavior, while their non-sonicated analogues show intercrystalline fracture behavior.

In this context, two processes accompanying sintering can be considered: grain boundary diffusion, which leads to an increase grain size, and grain boundary sliding, which leads to decrease a pore size. Ultrasonic treatment, which has an activation effect on the green bodies, makes it possible to intensify these processes by different ways during sintering.

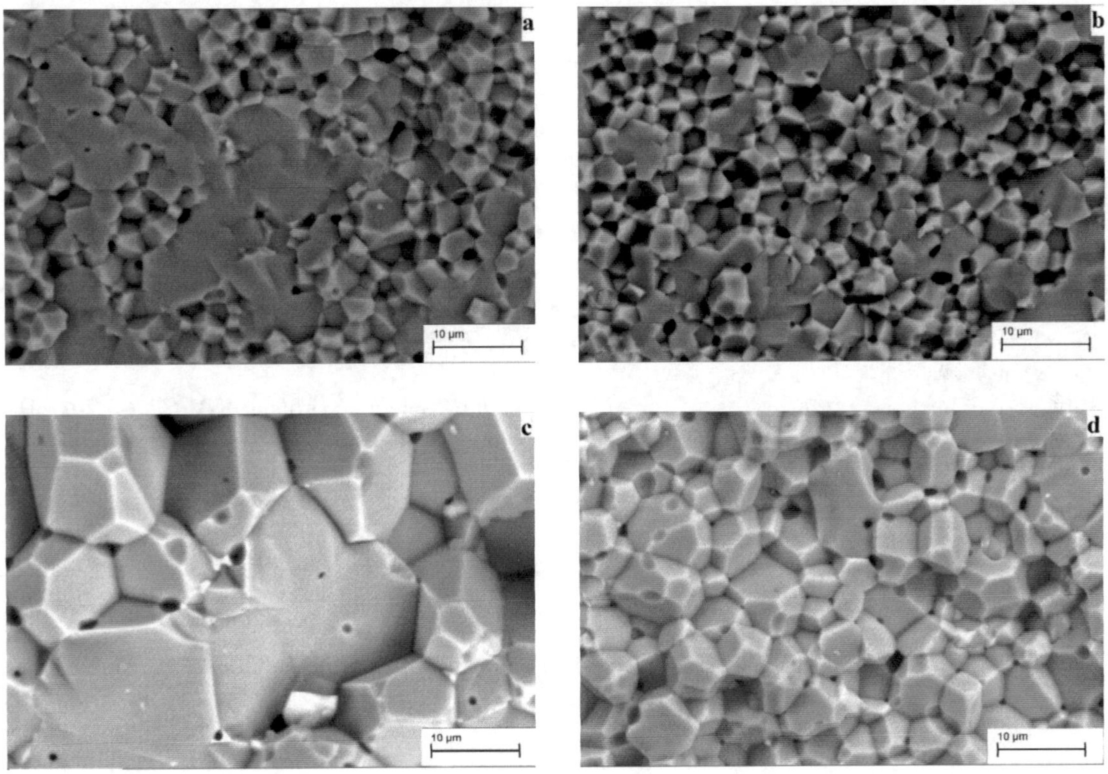

Figure 3. SEM-images of the fractured surface of ceramics sintered at (a) - 1650 °C; (b) - 1650 °C sonicated; (c) - 1700 °C; (d) - 1700 °C sonicated.

Thus, at sintering temperature - 1650 °C activation of the powder particle surface caused by ultrasonic treatment leads to a parallel intensification of two processes: grain boundary diffusion and grain boundary sliding, which increase grain sizes and decrease sizes of sonicated samples in comparison with non-sonicated samples. However, the density increase of sonicated samples sintered at this temperature indicates a slight predominance of grain boundary sliding.

Sintering at maximum temperature is accompanied by a significant acceleration of grain growth with a limited rate of pore size reduction, which indicates the dominance of the grains coalescence processes over the processes of their grain boundary sliding. In this case, the effects of closing porosity were observed. At sintering temperature - 1700 °C relative density, average grain and pore size of sonicated ceramic samples less than those of non-sonicated ceramic samples. Thus, the ultrasonic treatment effect on these samples, in contrast to samples sintered at 1650 °C, led to a parallel decrease in both pore sizes and grain sizes. This fact also confirms the assumption that ultrasonic activation intensify grain boundary sliding makes it possible to limit the diffusion coalescence of grains at high sintering temperatures. However, a slight density decrease of the sonicated samples indicates that the pore content increased. It may be due to the effect of closing porosity mentioned above.

Obviously, ultrasonic pressing at 400 MPa, followed by sintering at a temperature of 1650 ° C, is optimal condition among the investigated range, which can significantly reduce both the pore size and the value of residual porosity while maintaining grain sizes at an acceptable level.

X-ray phase analysis confirmed that ceramic samples consist of stoichiometric yttrium-aluminum garnet of cubic modification. All observed peaks are characteristic for the garnet phase.

The presence of other phases in the samples of ceramics were not detected. XRD reflections do not show any shifts in position or additional phases, indicating that the cerium were inserted completely in the structure.

4. Conclusion

High density ceramics based on Ce:YAG were produced by ultrasonic pressing at room temperature followed by conventional sintering on air. The effect of ultrasound treatment during the compaction process on density and microstructure of sintered ceramic has been studied.

Sintering condition was determined (Ultrasonic pressing at 400 MPa, sintering temperature - 1650 °C, isothermal exposure - 8 hours), which allows obtaining ceramics based on Ce: YAG with density about 100%, an average grain size - 4.57 μm and an average residual pore size - 0.94 μm.

It was found that the applied of ultrasonic treatment intensifies the process of grain growth, and in the case of manufacturing of ceramics in modes other than optimal it can lead to a decrease in density, the appearance and increase in the quantitative content of intragranular pores.

Acknowledgment

The work has been supported by government assignment "Science" project № 11.7700.2017/BCh Equipment of the Nano-center "Nanomaterials and Nanotechnologies" of National research Tomsk Polytechnic University has been used in the work.

References

[1] Lukin E S 1997 Modern high-density oxide ceramics with controlled microstructures. Part VI. Production of optically transparent oxide ceramic materials *Refractories and Industrial Ceramics* **7** 4-10

[2] Wang S F, Zhang J, Luo D W, Gu F, Tang D Y, Dong Z L, Tan G E B, Que W X, Zhang T S, Li S, Kong L B 2013 Transparent ceramics: processing, materials and applications *Progress in Solid State Chemistry* **41** 20-54

[3] Tang Y, Zhou S, Yi X, Hao D, Shao X, Chen J 2018 The characterization of Ce/Pr-doped YAG phosphor ceramic for the white LEDs *Journal of Alloys and Compounds* **745** 84-89

[4] Varney C R, Reda S M, Mackay D T, Rowe M C, Selim F A 2011 Strong visible and near infrared luminescence in undoped YAG single crystals *AIP Advances* **1.4** 042170

[5] Mihokova E, Nikl M, Mares J A, Beitlerova A, Vedda A, Nejezchleb K, Blazer K, D'Ambrosio C 2007 Luminescence and scintillation properties of YAG: Ce single crystal and optical ceramics *Journal of luminescence* **126** 77-80

[6] Singh G, Thomas V, Tiwari V S, Karnal A K 2017 Effect of cerium doping on optical and scintillation properties of transparent YAG ceramic *Ceramics International* **43** 9032-9040

[7] Dong J, Shirakawa A, Ueda K 2006 Efficient Yb^{3+}: $Y_3Al_5O_{12}$ ceramic microchip lasers 2018 *Applied physics letters* **89** 091114

IOP Conf. Series: Materials Science and Engineering **754** (2020) 012011 doi:10.1088/1757-899X/754/1/012011

[8] Valiev D, Han T, Stepanov S, Vaganov V and Paygin V 2018 The effect of BaF_2 concentration and particle size distribution on the luminescence efficiency of YAG: Ce^{3+} phosphors *Materials Research Express* **5** 096201

[9] Xia Z, and Andries M 2017 Ce^{3+}-Doped garnet phosphors: composition modification, luminescence properties and applications *Chemical Society Reviews* **46** 275-299

[10] Chen D and Yan C 2014 Transparent Ce^{3+}:$Y_3Al_5O_{12}$ glass ceramic for organic-resin-free white-light-emitting diodes *Ceramics International* **40** 15325-15329

[11] Osipov V V, Ishchenko A V, Shitov V A, Maksimov R N, Lukyashin K E, Platonov V V, Orlov A N, Osipov S N, Yagodin V V, Viktorov L V, Shulgin B V 2017 Fabrication, optical and scintillation properties of transparent YAG: Ce ceramics *Optical Materials* **71** 98-102

[12] Dvilis E S, Khasanov O L, Sokolov V M, Pokholkov Y P Method for compacting powder materials into articles and a mold for implementing the method *U.S. Patent No. 6919041 B2 19 Jul. 2005*

[13] Dvilis E S 2013 Regularities of the consolidation processes of powder systems at variable deformation and physical influences conditions Doctoral dissertation (In russian).

[14] Dvilis E S, Khasanov O L, Sokolov V M 2016 The relationship between the parameters of the state of pressing powder and properties of ceramics sintered from it *Fundamental Research* **11** 691-697

IOP Conf. Series: Materials Science and Engineering **754** (2020) 012012 doi:10.1088/1757-899X/754/1/012012

Effect of thermomechanical impacts on the deformation of thin polyimide films at uniaxial tension

A I Kupchishin[1,2], B G Taipova[1], M N Niyazov[1], V M Lisitsyn[3] and N A Voronova[1]

[1] Abay Kazakh National Pedagogical University, 13 Dostyk, Almaty, Kazakhstan

[2] Al-Farabi Kazakh National University, 71 al-Farabi ave., Almaty, Kazakhstan

[3] National Research Tomsk Polytechnic University, Tomsk, Russia

E-mail: ankupchishin@mail.ru

Abstract. Thermomechanical studies of the dependence of deformation (ε) on time (t) and temperature (T) for various static stresses in polyimide under uniaxial tension were carried out. Temporal deformation curves for static stresses in the range from 75 to 110 MPa consist of two stages: 1 – increase; 2 – exit to saturation, which is associated with a change in the nature of structural transformations of macromolecules, due to the action of the static stresses. An increase in temperature and voltage leads to an increase in the dependence ε (T). This is due to the intensification of the processes of structuring macromolecules with increasing temperature and are supplemented by processes of breaking bonds between macromolecules, which increase the strain rate. Experimental data are described in terms of exponential and linear models.

1. Introduction

The most close attention among polymer systems of various classes is caused by materials based on aromatic polyimides (PI) [1, 2], which have a unique combination of extremely high heat and heat resistance, excellent mechanical and electrical characteristics combined with high chemical and radiation resistance [3, 4].

PI matrices retain the ability to crystallize after melting and subsequent cooling [5, 6]. In this regard, it is relevant to obtain new fundamental results in the field of deformation and thermomechanical properties of materials, to determine the safe strength life of products under the combined effect of external loads and operating parameters, and to assess the hazard of technological and temporal factors [7 – 11].

Polyimides, including the alicyclic structure, are the most important polymers for the successful solution of problems in materials science. In this area, research is carried out in two directions – along the path to creating polymers with electrical insulating and electrically conductive properties. The improvement of individual physicomechanical properties leads to an improvement in the characteristics of polyacrylates, in particular, elasticity, which also expands the field of application of alicyclic polyimides [4].

In [12, 13], it was found that polyfunctional amines, in particular tetraamines, bis (o-phenylenediamines) and bis- [3-amino-4 (p-aminophenoxy)] arylenes, are widely used to modify PIs. When polyanilines are used as polyfunctional amines, PIs with good performance properties associated with the formation of block film-forming materials of spatial structure were obtained [14].

The effect of moisture on charge relaxation processes in modified polyimide films with one-sided and two-sided teflon coatings was established in [15].

This paper is devoted to the study of the complex effects of various static loads and temperatures on the mechanical properties of polyimide.

2. Experimental

Industrial polyimide with a thickness of 45 μm was selected as the test material. Film samples were cut using a special device. The length of the test material was 70 mm, the working part 50 mm, width 5 mm.

Studies of the dependence of ε on *t* carried out on the developed experimental setup, consisting of the following blocks: case, bracket, upper clamping device, sample under study, lower clamp, displacement sensor, load system, force (voltage) sensor, and video camera. The body is a tripod with a variety of mounts, on which all parts of the installation are located. When measuring the relative elongation (ε) and stress (σ), a special interface with various measurement sensors (Science Cube) was used. To conduct experimental studies, we determined the maximum load with the help of special cargoes, i.e. the maximum value of the force at which the sample was torn. For this material $\sigma = 112$ MPa. Then the voltage was calculated, which constitutes a certain part of the limit voltage. After that, the sample was fixed in clamps, and the dependence of ε on *t* was recorded. The time of testing the sample for uniaxial tension was about 50 seconds. The data was recorded on a surveillance camera for further processing.

When conducting research on the dependence of ε on temperature (T), an installation has been designed and manufactured that makes it possible to measure parameters using motion and force sensors at different loads and their change with time. Figure 1 shows the temperature as a function of time in the upper part of the installation when it is heated to 180 °C. It was obtained that the error lies within 5 %.

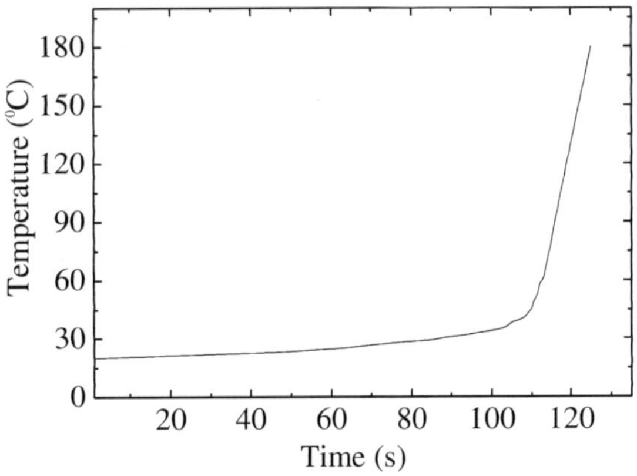

Figure 1. Dependence of temperature on time in the upper part of the temperature unit when it is heated to 180 °C

3. Results and discussion

A series of experiments was carried out to study the dependence of ε on t under various static loads. The graphs are presented in Figure 2. The patterns of strain change allow to study and evaluate the nature of the structural changes due to the effect of static load. Exposure to a polyimide film with a voltage of 60 MPa does not lead to a change in the relative elongation (Figure 2, curve 1).

The deformation-time curves for static loads in the range of 75 – 110 MPa have a similar character, consisting of two stages: 1 – in the range of 0 – 12 seconds, the relative elongation of the material

IOP Conf. Series: Materials Science and Engineering **754** (2020) 012012 doi:10.1088/1757-899X/754/1/012012

increases; 2 – at t = 12 – 50 seconds ε comes to saturation. Curve 1 is described by a linear function ε = at (a = 0), and curves 2 – 5 by the formula:

$$\varepsilon = \varepsilon_0 \left(1 - \exp\left(-\frac{t}{t_0}\right)\right).$$

The values of ε_0 and t_0 are listed in the table.

Table 1 – Dependence of ε_0 and t_0 on σ

σ, [MPa]	75	85	100	110
ε_0, [%]	29	34	42	47
t_0, [sec.]	4	5	7	8

It is seen that with increasing σ, ε_0 and t_0 grow.

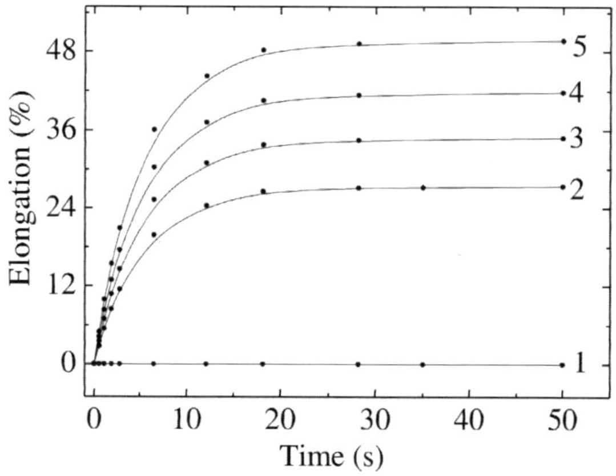

1 – 60; 2 – 75; 3 – 85; 4 – 100; 5 – 110 MPa;
points – experiment; solid lines – calculation

Figure 2. Dependence of deformation on the time of the polyimide film at various static stresses

Then, the dependences of ε on T were studied experimentally for various values of the static loads σ_1 = 60; σ_2 = 75; σ_3 = 85; σ_4 = 100 and σ_5 = 110 MPa, shown in Figure 3. As can be seen from the figure, the relative elongation increases with increasing T and σ. As the temperature rises, the destruction processes are intensified.

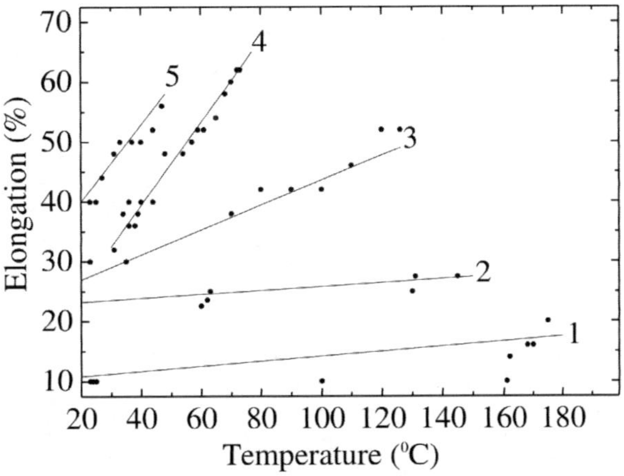

1 – 60; 2 – 75; 3 – 85; 4 – 100; 5 – 110 MPa
Points – experiment; solid lines – calculation according to Hooke

Figure 3. The dependence of the relative elongation of the temperature of the polyimide film at different static loads

The first three curves indicate the intensification of the processes of structuring macromolecules with increasing temperature, and curves 4 and 5 indicate a change in the nature of the transformation of the polymer structure, which are complemented by the processes of breaking the bonds between macromolecules, which increase the deformation. The calculation according to [16] leads to a change in the tangent of the angle of inclination from 0.04 to 0.64.

The dependence of the maximum relative elongation (ε_{max}) on the static load of the polyimide film during thermomechanical testing is presented in Figure 4. The analysis shows that this dependence is characterized by a monotonic increase with saturation. Such behavior of ε indicates a change in the nature of the structural transformations of macromolecules, due to the specificity of the action of static load and temperature. The dependence of ε on σ is described by the expression:

$$\varepsilon_{\max} = \varepsilon_0\left(1 - \exp\left(-\frac{\sigma}{\sigma_0}\right)\right),$$

where for this case $\varepsilon_0 = 57$ % and $\sigma_0 = 60$ MPa.

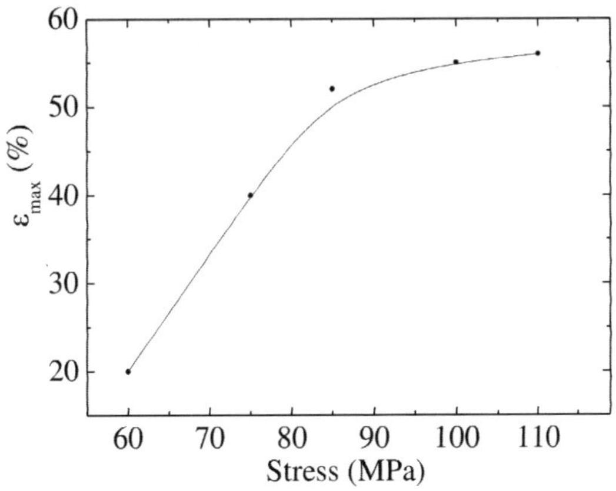

Figure 4. Dependence of maximum relative elongation of polyimide film on static load during thermomechanical testing

Further, the dependence of the residual deformation ε_{res} on σ was investigated (Figure 5). As can be seen from figure 5, the dependence is linear.

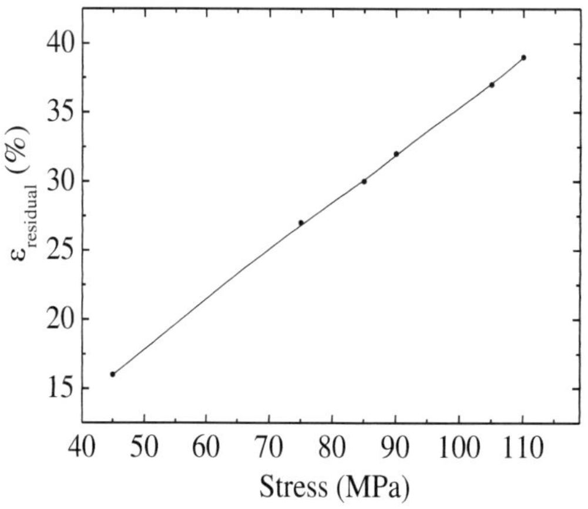

Figure 5. Dependence of the residual deformation on the static load of the polyimide film during thermomechanical testing

4. Conclusion

1. Experimental work has been carried out to study the dependence of deformation on time for various values of static stresses. The deformation-time curves for static stresses in the range from 75 to 110 MPa have a similar character, consisting of two stages: 1 – in the range of 0 – 12 seconds, the relative elongation of the material increases; 2 – at t = 12 – 50 sec it reaches saturation. Such behavior indicates the change in the nature of the structural transformations of macromolecules due to the action of the static stresses.

2. It was found that the dependence of ε on T increases with both increasing T and σ. This is due to the intensification of structuring of macromolecules with increasing temperature and it is complemented by the processes of breaking bonds between macromolecules, which increase the rate of deformation.

References

[1] Krutko E T, Prokopchuk N R and Globa A I 2017 Chemical modification of polypyromellimtimide *Polymeric materials and technologies* **3** 1 33-46

[2] Liu B, Zhang B, Wang X, Chen X and Gu G 2019 The preparation and properties of polyimide films modified by octa(aminopropylsilsesquioxane) *Journal of Saudi Chemical Society* (In press. https://doi.org/10.1016/j.jscs.2019.02.002)

[3] Zuev V V, Uspenskaya M V and Olekhnovich A O 2010 *Physics and chemistry of polymers Textbook* (SPb.: SPbSU ITMO)

[4] Kudaykulova S K, Iskakov R M, Kravtsova V D, Umerzakova M B, Abadie M, Zhubanov B A at al 2006 *Polymers for special purposes* (Almaty)

[5] Lian M, Zheng F, Lu X and Lu Q 2019 Tuning the heat resistance properties of polyimides by intermolecular interaction strengthening for flexible substrate application *Polymer* **173** 205-214

IOP Conf. Series: Materials Science and Engineering **754** (2020) 012012 doi:10.1088/1757-899X/754/1/012012

[6] Kim K, Yoo T, Kim J and Han H 2015 Effects of dianhydrides on the thermal behavior of linear and crosslinked polyimides *J.Appl. Polym. Sci.* **132** 6 44412-21

[7] Gerasimenko N N, Voronova N A, Kovtunets V A, Taipova B G, Kupchishin A I and Omarbekova Z A 2003 Study of the mechanical properties of polymer films effecting on their by static load and temperature *Proceedings of the universities. Materials of electronic technology* 6-12

[8] Sh.-Sh Dong, W.-Zh Shao, Li Y, Hui-Jian Ye and Liang Zhen 2018 Microstructure evolution of polyimide films induced by electron beam irradiation-load coupling treatment *Polymer Degradation and Stability* **155** 230-237

[9] Rahnamoun A, Engelhart D P, Humagain S, Koerner H, Plis E, Kennedy W J, R Cooper, Greenbaum S G, Hoffmann R and Adri CT van Duin 2019 Chemical dynamics characteristics of Kapton polyimide damaged by electron beam irradiation *Polymer* **176** 135-145

[10] Sorokin S V, Terentyev A V and Krotov E A 2014 *Introduction to the mechanics of a deformable solid* (SPb: SPbGMTU)

[11] Kupchishin A I, Niyazov M N, Voronova N A, Kirdiashkin V I and Abdukhairova A T 2017 The effect of temperature, static load and electron beam irradiation on the deformation of linear polymers *IOP Conf. Series: Materials Science and Engineering 168012016* 1 – 4

[12] Babman T, Hamid Y 2001 Preparation and properties of novel polyimides de-rived from 4-aril-2,6-bis(4-aminophenyl)pyridine *J. Polym. Sci. Part A, Polym. Chem.* **39** 21 3826-31

[13] Lee D H et al. 2002 Preparation and properties of aromatic polyimides based on 4,4-(2,2,2-trifluoro-1-phenylethylidene)diphthalic anhydride *J. Appl. Polym. Sci.* **85** 1 38-44

[14] Meleshko T K et al. 2009 Electrically conductive film-forming compositions based on a mixture of polyaniline and polyimides *Macromolecular compounds. Ser. A* **51** 3 447-453

[15] Kamalov A M and Borisova M E 2016 The effect of moisture on the relaxation of electric charge in modified films based on polyimide *Scientific-technical statements SPbGPU. Physics and Mathematics* **3** (248) 33-38

[16] Voronova N A, Kupchishin A I, Niyazov M N and Lisitsyn V M 2018 Uniaxial Stress and Electron Irradiation Effects on Nanochains Straightening in Film Polymer Materials *Key Engineering Materials* **769** 78-83

IOP Conf. Series: Materials Science and Engineering **754** (2020) 012013 doi:10.1088/1757-899X/754/1/012013

Electrical discharge machining of polymer composite materials

T Ablyaz[1], K Muratov[1], E Shlykov[1], E Morozov[1], V Karmanov[1], T Shakirzianov[1], S S Sarabjeet[2] and D Sahil[3]

[1] Perm National Research Polytechnic University, Perm, 614000, Russian Federation
[2] Mechanical Engineering Department, Beant College of Engineering & Technology, Gurdaspur-143521, India
[3]Thapar Institute of Engineering and Technology, Patiala, 147001, India

*E-mail: lowrider11-13-11@mail.ru

Abstract. Improving the technical capabilities of manufacturing processes and creating new parts that surpass existing quality benchmarks is a need of modern engineering. Polymer Composite Materials (PCM) are commonly used in the manufacturing of advanced aerospace components. The aim of this work is to study the process of Electrical Discharge Machining (EDM) of PCM VKU-29. To study the process performance, three different types of Electrode Tools (ET) of different materials were used. EDM was performed on three modes viz. maximum, medium, and minimum. The influence of process parameters and the electrode type on the formation of spark gap was investigated. The visual and the microscopic inspection revealed the absence of hairiness and traces of fiber rupture in case of composite ET. Thus, it can be concluded that the combination of smallest value of spark gap and the composite ET leads to the best quality of surface amongst all the other combinations. Moreover, while processing the PCM VKU-29 using composite ET, it is advisable to maintain a spark gap of 0.14 mm for the best results.

1. Introduction

Improving manufacturing capabilities and creating new components with their properties that are comparable or better to the existing is an imperative challenge for modern engineering. Polymer Composite Materials (PCM) are commonly used in the manufacturing of advanced aerospace components. PCMs are a kind of versatile and high performance material formed from a combination of two or more phases of materials [1]. Structural PCMs are currently being developed with high resistance to static and dynamic loads [2]. These materials have special properties, with a wide temperature range of operation. Due to the use of different binders, these materials are able to provide properties like heat resistance and strength. These properties ensure the machining of such structures [3, 4]. PCMs also have the ability to effectively damp the background noise associated with the operation of turbine blades used in aircrafts. Thus, noise absorption panels of aircrafts are usually made of these materials [5].

However, these materials have many advantages than the conventional materials but the processing of these materials is difficult. The manufacturing of components from this material has some limitations [6]. Yaroslavtsev et al. studied the PCM machining and concluded that on machining, the wear of the cutting tool increases which leads to the decrease in productivity [7]. Also, due to very low stiffness, it is difficult to machine these materials with multipoint cutting tools and abrasive blades.

During machining (drilling, turning, milling and surface machining) of PCM components, defects such as peeling and tearing of the fiber from the matrix arise. Surface defects such as low and uneven roughness, chips at the ends, cracks, and fiber delamination can be attributed to the cutting action of the tool on the two phased material [1, 2]. Thus, there is a demanding need of the research and development of such alternative machining processes which can machine materials like PCM with high accuracy [8-10]. An adequate solution to this can be the use of electrical discharge machining (EDM). Electrical discharge machining (EDM) is a non-conventional method of machining focused on removing material from a component through a sequence of periodic electrical discharges between electrodes, and the part being machined in the presence of a dielectric fluid [11]. The processes taking place during the electric discharge are presented in Figure 1.

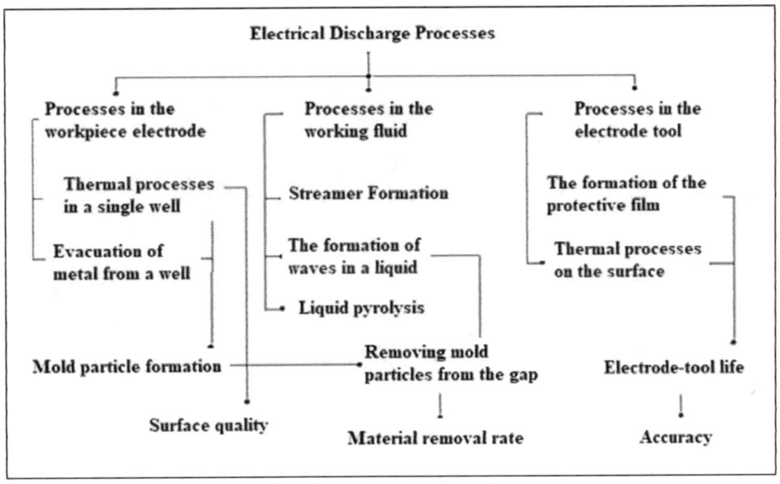

Figure 1. The basic physical processes during EDM.

EDM of a variety of conductive materials have been investigated in the literature. But there is a dip in the literature for EDM of non-conducting materials. Specifically, EDM of PCM is not completely studied. Thus the aim of this current work is to study the process of EDM of PCM VKU-29.

2. Materials and Methods

For the current work, composite material- carbon fiber VKU-29 was selected. EDM of VKU-29 was carried out by three different Electrode Tools (ET). The details of the ET used have been given in the **Table 1**. The machining was carried out in three modes with a constant pulse duration (Ton) = 150 μs and open voltage (U) = 100 V. The value of current (I) was varied as 8A (maximum), 2A (medium), and 0.5 A (minimum). The machined surface and the ET after the machining were analyzed under Carl Zeiss CONTURA G2® coordinate measuring machine. An average of three readings were taken as the final value to eliminate any discrepancy during the analysis.

Table 1. Different types of ET used in this study.

S.No.	Electrode Tool	Standard
1	Graphite	GOST 4426-80

IOP Conf. Series: Materials Science and Engineering **754** (2020) 012013 doi:10.1088/1757-899X/754/1/012013

2	Composite Material	(Pseudo alloy of the copper + colloidal graphite (CuC) system) Powders PMS-1 (GOST 49-60-75) and dry colloidal graphite of the S-1 grade (TU 113-08-48)
3	Copper (M1 grade)	GOST 1173–2006

3. Results and discussion

The burning of the samples have been observed during the maximum conditions (Maximum EDM mode). Figure 2 presents the values of the spark gap, measured as the difference between the width of the machined surface and the width of the ET before processing.

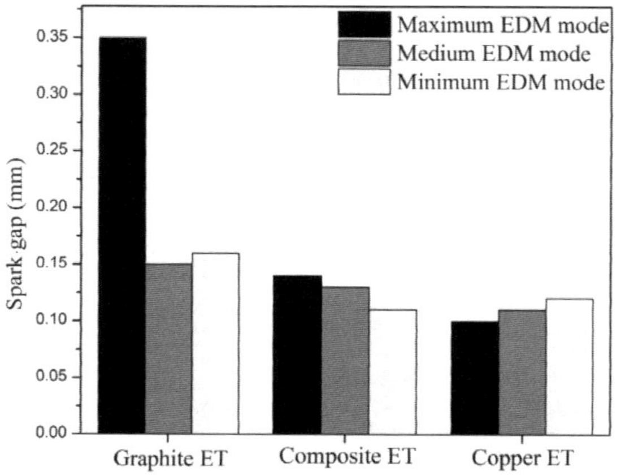

Figure 2. The value of spark gap during EDM of VKU-29 with three different ETs.

The results of the study shows that it is difficult to machine the PCM with graphite ET in all modes. The value of spark gap during machining at maximum mode is the largest. Composite ET processing is characterized by the best quality of machined surface, and has a smaller value of the spark gap at all processing modes (Spark gap = 0.23-0.30 mm). When processing copper Electrode-tool in various modes, the spark gap is minimal (Spark gap = 0.23-0.32 mm), however, the surface of the VKU-29 thermoplastic was found to be fleecy. As a result of the experiments, it has been concluded that EDM of PCM by composite ET provides the best quality of surface (Figure 3).

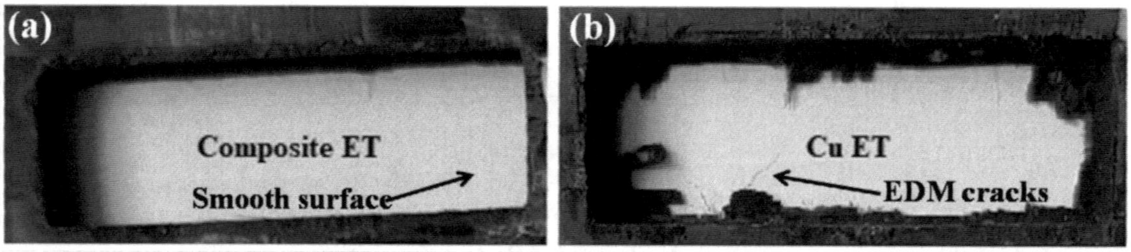

Figure 3. Surface characteristics of VKU-29 after EDM of (a) composite ET and (b) Copper ET.

It was found that during EDM with composite ET, the treated surface has no traces of torn layers and hairiness, and subsequent finishing is not required. The effect of "sintering" of PCM fiber on the treated surface was observed. The machining process was accompanied by stable sparking without growths on the working surfaces of ET. In a similar study it has been shown that it is entirely feasible to machine carbon fibre composite materials by EDM process. The advantage of such a process with carbon fibre composite materials is the capability of producing irregular shaped holes with good surface finish and

IOP Conf. Series: Materials Science and Engineering **754** (2020) 012013 doi:10.1088/1757-899X/754/1/012013

dimensional accuracy. Also, Copper electrodes prove to be better than graphite electrodes in terms of tool wear and surface finish [12].

4. Conclusions

The absence of hairiness and traces of fiber rupture on the machined surface concludes that it is possible to use EDM technology with composite ET for machining PCM VKU-29. It has been shown that while using composite ET for the EDM of VKU-29 material, it is necessary to maintain a spark gap of 0.14 mm for the best surface.

Acknowledgement
Financial support was provided by the Russian President in encouragement of young Russian scientists (grant MK-2072.2019.8).

References
[1] Rahman R and S. Zhafer Firdaus Syed Putra, "5 - Tensile properties of natural and synthetic fiber-reinforced polymer composites," in *Woodhead Publishing Series in Composites Science and Engineering*, Woodhead Publishing, 2019, 81–102
[2] Zlobina I and Bekrenev N, "Interaction of microwave electromagnetic field with constructions made of reinforced carbon plastic with metal cellular elements inbuilt," *Mater. Today Proc.*, 2019
[3] Bains P S, Sidhu S S, and Payal H S 2016 "Fabrication and Machining of Metal Matrix Composites: A Review," *Mater. Manuf. Process.*, **31** **5** 553–573
[4] Bains P S, Singh S, Sidhu S S, Kaur S, and Ablyaz T R, "Investigation of Surface Properties of Al–SiC Composites in Hybrid Electrical Discharge Machining BT - Futuristic Composites : Behavior, Characterization, and Manufacturing," Singapore: Springer Singapore, 2018, 181–196
[5] Rychkov D A, Yanyushkin A S, Lobanov D V, Bazarkina V V 2012 "Improving the technology of forming high-strength fiberglass composite materials on a polymer basis" Metal processing (technology, equipment, tools) **3** 150-153
[6] Kablov E N 2012 "Strategic directions for the development of materials and technologies for their processing for the period up to 2030" Aviation materials and technologies: Anniversary scientific and technical. Sat (Appendix to the journal "Aviation materials and technologies"). M .: VIAM, 7-17
[7] Ablyaz T R, Shlykov E S, and Kremlev S S 2017 "Copper-coated electrodes for electrical discharge machining of 38X2H2MA steel," *Russ. Eng. Res.*, **37** **10** 910–911
[8] Sidorov D and Kramar V 2019 "Modelling of Holes Shaping at Drilling of Polymer Composite Materials," *Mater. Today Proc.* **11** 600–607
[9] Palanikumar K and Davim J P 2013 "5 - Electrical discharge machining: study on machining characteristics of WC/Co composites," in *Woodhead Publishing Reviews: Mechanical Engineering Series*, Woodhead Publishing, 135–168
[10] Sidhu S S, Batish A, and Kumar S 2013 "Fabrication and electrical discharge machining of metal-matrix composites: A review," *J. Reinf. Plast. Compos.* **32** **17** 1310–20
[11] Sidhu S S, Batish A, and Kumar S 2014 "ED Machining of Particulate Reinforced MMC ' s," *Int. J. Mech. Aerospace, Ind. Mechatronics Eng.* **8** **3** 519–525
[12] Lau W S, Wang M, and Lee W B 1990 "Electrical discharge machining of carbon fibre composite materials," *Int. J. Mach. Tools Manuf.* **30** **2** 297–308

IOP Conf. Series: Materials Science and Engineering **754** (2020) 012014 doi:10.1088/1757-899X/754/1/012014

Luminescence spectra of YAG:Ce phosphors synthesized in a field of radiation

G Alpyssova[1], D Mussakhanov[1,2], Zh Karipbayev[1,2], T Grechkina[2], Shiming Zheng[2] and A Kukenova[1]

[1]L.N. Gumilyov Eurasian National University, Nur-Sultan, Kazakhstan
[2]National Research Tomsk Polytechnic University, Tomsk, Russian Federation

E-mail: gulnur-0909@mail.ru

Abstract. The paper presents the measurement results of the luminescence band shape for YAG phosphors upon photoexcitation. It is shown that the position and FWHM does not depend on the excitation technique, but it depends on the phosphor prehistory. The observed difference is assumed to be due to the difference in the structure of nanodefects in phosphors synthesized under different conditions.

1. Introduction

A large number of studies are aimed to develop new synthesis technologies and new compositions to increase the efficiency of phosphors for LEDs. YAG:Ce based phosphors are currently considered the most promising, and many papers are devoted to investigation of these phosphors. The analysis of the published data on spectral characteristics draw attention to distinctions in FWHM and the position of the luminescence bands of phosphors synthesized under different conditions [1-4]. We performed studies of YAG:Ce phosphors of different prehistory to find out the reasons for these distinctions.

2. Experimental

The luminescence spectra were measured for LED lamps and YAG phosphors synthesized at NPO Platan (Russia) and Fultor Enterprises Co. Ltd. (China), and phosphors synthesized in the field of the radiation flux upon excitation by the chip with λ=450 nm and nitrogen laser with λ=337 nm. The study employed SDL3500, SDL4000, YAG 01 and YAG 02 industrial phosphors that obviously differ in technological synthesis modes. The rest of the phosphors were synthesized in the field of the radiation flux [5-7]. Phosphors were obtained by mechanical crushing synthesis and annealed at a temperature of at 1650 °C for 8 hours. After that luminescence spectra of phosphors were measured.

To achieve high measurement accuracy for characteristics of the luminescence bands, a setup was assembled. Figure 1 shows the block diagram of the setup.

A chip with λ=450 nm and a nitrogen laser with λ = 337 nm were used as excitation sources. Both excitation sources excited YAG:Ce phosphors at the maxima of the phosphor absorption bands. The excitation flux was directed to the phosphor. A series of the studied phosphors was placed on a glass substrate in metal washers 1 mm thick with an inner diameter of 6.5 mm.

IOP Conf. Series: Materials Science and Engineering **754** (2020) 012014 doi:10.1088/1757-899X/754/1/012014

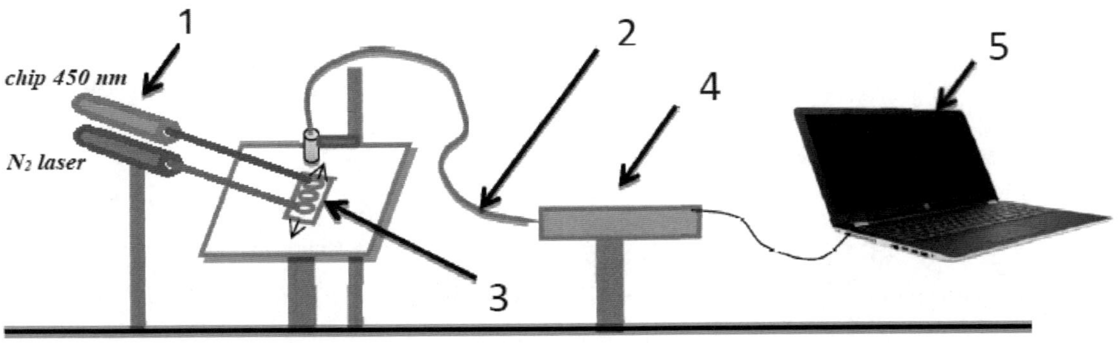

Figure 1. Scheme of the setup to measure the luminescence spectra: excitation sources (1), optical fiber (2), samples on the substrate (3), Avantes AvaSpec-2048L spectrophotometer (4), notebook (5)

The excited luminescence was transmitted through a quartz optical fiber to an Avantes AvaSpec2048L spectrophotometer. The position of the input window of the light guide did not change relative to the sample with the phosphor. A spectrophotometer records the luminescence spectrum, which was then displayed on the laptop display for further processing. The mutual arrangement of all the setup components: the excitation source, the washers with the phosphor, the LED window remained unchanged during the experiment. The measurements were carried out under the same conditions.

During the measurement, the mutual arrangement of the circuit components and the power supply was chosen so as not to distort the shape of emission bands. The position of the band and its FWHM were determined using the WebPlotDigitizer [7] software. The type of data displayed on the monitor is shown in figure 2 as an example.

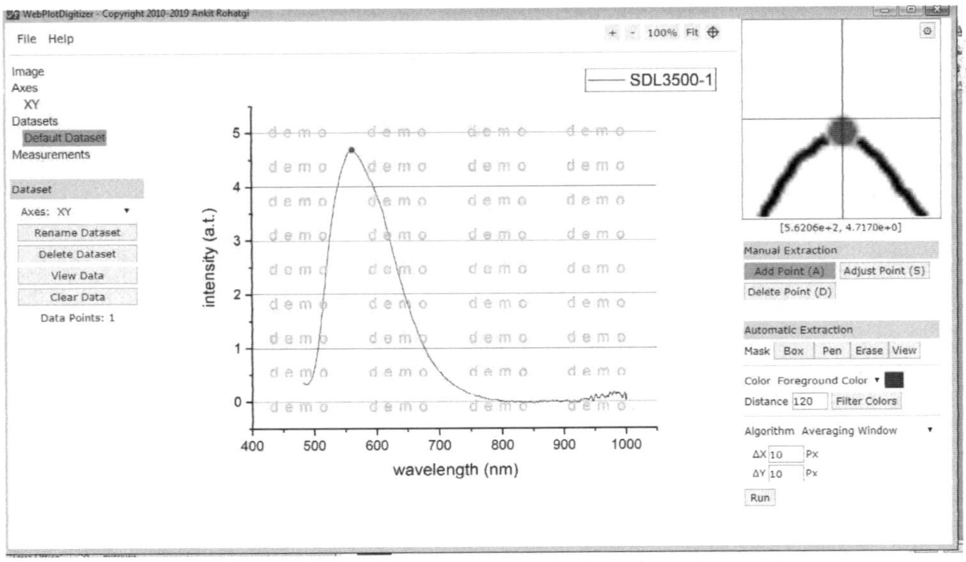

Figure 2. Data in WebPlotDigitizer displayed on the monitor.

3. Results and discussion

In total, 24 types of phosphors were chosen for research. Among them, 4 types of industrial phosphors: SDL3500, SDL4000, YAG 01, YAG 02. The elemental composition of phosphors is summarized in table 1. It was determined using the SEM Quanta3D 200i with XRD system.

Table 1. Elemental composition of the samples of the studied phosphors, in at%

Phosphor	Y	Al	O	Gd	Ce
SDL 3500	13.8	26.0	57.3	2.5	0.4
SDL 4000	13.8	27.0	58.8		0.4
YAG-01	12.5	29.4	54.0	2.7	1.4
YAG-02	12.9	28.0	58.2		0.9

Twenty phosphors were made from ceramics synthesized in the field flow of electron [8]. Luminescence was measured for 10 samples after synthesis and crushing and for 10 samples after subsequent thermal annealing of the same samples at 1650 °C in vacuum. The compositions of 10 ceramic samples are presented in table 2.

Table 2. Elemental composition of the synthesized phosphors

#	Samples before annealing
1	$Y_3Al_5O_{12}$: Ce_2O_3(4.8%)
2	$Y_3Al_5O_{12}$:Ce_2O_3(2%)
3	$Y_3Al_5O_{12}$:Ce_2O_3(2%)
4	$Y_3Al_5O_{12}$:Ce_2O_3(2%)
5	$Y_3Al_5O_{12}$:Ce_2O_3(2%) + Gd_2O_3(6%)
6	$Y_3Al_5O_{12}$:Ce_2O_3(2%) + Gd_2O_3(6%)
7	$Y_3Al_5O_{12}$:Ce_2O_3(2%) + Gd_2O_3(6%)
8	$Y_3Al_5O_{12}$:Ce_2O_3(2%)
9	$Y_3Al_5O_{12}$:Ce_2O_3(4%)
10	$Y_3Al_5O_{12}$: Ce_2O_3(1%) + Gd_2O_3(13%)
#	**Samples after annealing**
1	$Y_3Al_5O_{12}$:Ce_2O_3(4.8%)
2	$Y_3Al_5O_{12}$:Ce_2O_3(2%)
3	$Y_3Al_5O_{12}$: Ce_2O_3(2%)
4	$Y_3Al_5O_{12}$:Ce_2O_3(2%)
5	$Y_3Al_5O_{12}$:Ce_2O_3(2%) + Gd_2O_3(6%)
6	$Y_3Al_5O_{12}$:Ce_2O_3(2%) + Gd_2O_3(6%)
7	$Y_3Al_5O_{12}$:Ce_2O_3(2%) + Gd_2O_3(6%)
8	$Y_3Al_5O_{12}$:Ce_2O_3(2%)
9	$Y_3Al_5O_{12}$:Ce_2O_3(4%)
10	$Y_3Al_5O_{12}$: Ce_2O_3(4.8%) + Gd_2O_3(11.9%)

Phosphors differed in the presence of Ce_2O_3 and gadolinium ions incorpotrated as a modifier. Phosphors 2–4 and 5–7 differed in the prehistory of the starting alumina. The used Al_2O_3 powders were produced in Labor Pharma LLP, Kazakhstan (ChDA TU 6-09-426-75 qualification), Chemical Reagents Plant, Russia (ChDA TU 6-09-426-75) and Hefei Zhonghang Nanotechnology Development Co., Ltd., China (ZH-Al_2O_3-01).

For each type of phosphor, the spectra were measured 10 times, and the statistical measurement error in each case was calculated. The spectrum measurement results: position, FWHM, measurement error are presented in Table 3.

Table 3. Position and FWHM of the luminescence bands in phosphors

#	Type of phosphor	λ_{ex}=337nm		λ_{ex}=450nm	
		λ_{Max}, nm	ΔE, eV	λ_{Max}, nm	ΔE, eV
		Samples before annealing			
1	1	557±2	0.452±0.005	559±2	0.451±0.005
2	2	538±2	0.446±0.005	537±2	0.455±0.005
3	3	535±2	0.448±0.005	534±2	0.451±0.005
4	4	549±2	0.46±0.005	555±2	0.462±0.005
5	5	548±2	0.452±0.005	552±2	0.448±0.005
6	6	558±2	0.46±0.005	558±2	0.464±0.005
7	7	549±2	0.456±0.005	554±2	0.459±0.005
8	8	557±2	0.449±0.005	558±2	0.450±0.005
9	9	538±2	0.435±0.005	538±2	0.443±0.005
10	10	568±2	0.459±0.005	567±2	0.455±0.005
		Samples after annealing			
11	1	558±2	0.45±0.005	559±2	0.44±0.005
12	2	541±2	0.438±0.005	540±2	0.442±0.005
13	3	538±2	0.455±0.005	535±2	0.445±0.005
14	4	557±2	0.449±0.005	555±2	0.444±0.005
15	5	554±2	0.457±0.005	552±2	0.452±0.005
16	6	556±2	0.469±0.005	557±2	0.463±0.005
17	7	555±2	0.453±0.005	541±2	0.441±0.005
18	8	554±2	0.445±0.005	553±2	0.439±0.005
19	9	546±2	0.437±0.005	544±2	0.440±0.005
20	10	566±2	0.479±0.005	567±2	0.472±0.005
		Industrial phosphors			

21	YAG-01	562±2	0.438±0.005	562±2	0.438±0.005
22	YAG-02	558±2	0.438±0.005	558±2	0.441±0.005
23	SDL-3500	561±2	0.455±0.005	561±2	0.440±0.005
24	SDL-4000	557±2	0.441±0.005	560±2	0.446±0.005

The results obtained show that in all the measured phosphors the band position falls in the range from 535 to 568 nm, and FWHM is 0.435–0.479 nm. For the phosphors synthesized, the band position varies from 535 to 568 nm and FWHM is 0.435–0.479 nm; for industrial phosphors, the band position ranges from 558 to 562 nm, and FWHM is 0.438–0.455 nm.

4. Conclusion

The results obtained in the study of the band shape of phosphors of different prehistory allow us to draw the following conclusions. The positions of the luminescence bands and their FWHM depend on the prehistory, but not on the excitation technique. The difference in the position and FWHM of the bands can be partially attributed to the known effect of the presence of gadolinium. However, the dependence of the band shape in the samples before and after annealing cannot be explained by this effect. Annealing does not change the composition, but it causes ordering of the crystal structure, the environment of the luminescence centers, and cerium ions.

We believe that the difference in the band shape indicates that the structure of the environment of the luminescence centers in YAG:Ce samples of different prehistory is different. This can be due to different technological modes of their synthesis. It is assumed that the observed spread in the luminescence characteristics is caused by the difference in the structure of nanodefects in phosphors synthesized under different conditions [9]. The elemental composition in all nanodefects in YAG: Ce phosphors is similar: matrix ions, dopant, modifier, and intrinsic lattice defects, but their ratio and mutual distribution in the nanodefect is different.

Acknowledgement

This research was supported by Tomsk Polytechnic University CE Program.

References

[1] Lisitsyn V, Lisitsyna L and Polisadova E 2019 *Bulletin of the Russian Academy of Sciences: Physics* **83** 336–342

[2] Hongling Shi, Chen Zhu, Jiquan Huang, Jian Chen, Dongchuan Chen,Wenchao Wang, Fangyu Wang, Yongge Cao and Xuanyi Yuan 2014 *Optical materials express* **4** 649-655

[3] Kyeong Youl Jung and Hyun Woo 2007 *Journal of Luminescence* **126** 469–474

[4] Heesun Yang and Yong-Seog Kim 2008 *Journal of Luminescence* **128** 1570–76

[5] Lisitsyn V, Lisitsyna L, Dauletbekova A, Golkovskii M, Karipbayev Zh, D. Musakhanov, Akilbekov A, Zdorovets M, Kozlovskiy A and Polisadova E 2018 *Nuclear Inst. and Methods in Physics Research* B **435** 263–267

[6] Lisitsyn V, Golkovskii M, Lisitsyna L, Dauletbekova A, Musakhanov D, Vaganov V, Tulegenova A, and Karipbayev Zh 2019 *Russian Physics Journal* **61** 1909-13

[7] https://automeris.io/WebPlotDigitizer/

[8] Lisitsyn V, Golkovsky M, Musakhanov D, Tulegenova A, Abdullin Kh and Aitzhanov M 2018 *IOP Conf. Series: Journal of Physics: Conf. Series* 1115 052007

[9] Lisitsyn V, Lisitsyna L, Tulegenova A, Yangyang Ju. Polisadova E, Lipatov E and Vaganov V 2019 *Crystal* **9** 476

Effect of electronic irradiation on the properties of polyethyleneterephthalate films of various producers

A I Kupchishin[1,2], B G Taipova[1], N N Gerasimenko[3], N A Voronova[1] and AT Abdukhairova[1]
[1] Abay Kazakh National Pedagogical University, Almaty, Kazakhstan
[2] Al-Farabi Kazakh National University, Almaty, Kazakhstan
[3] National Research University Moscow Institute of Electronic Technology, Moscow, Russia

E-mail: ankupchishin@mail.ru

Abstract. Studies on the effect of electron irradiation on the properties of polyethylene terephthalate films of various brands have been carried out. An analysis of the results shows that the relative elongation of the Mylar-type film is ~ 70%, for PET ε it is 48%. According to the strength properties, Russian-made PET is more resistant to electronic effects and is destroyed at doses of 10^9 Gy, while the Mylar ones at 10^5 Gy. The calculation made according to the proposed models (exponential and generalized) is in satisfactory agreement with experimental data.

1. Introduction

Most of the polyethylene terephthalate (PET) in the world goes to the production of films and fibers. Its widest application is observed in mechanical engineering, the chemical industry, food equipment, transport and conveyor technologies, medical industry, instrument making and household appliances [1 – 3].

Polyethylene terephthalate is a common thermoplastic, a representative of the class of polyesters, known under different brand names. For physical properties, it is a white solid, odorless. PET is durable, tough and lightweight material. In Russia, polyethylene terephthalate is called lavsan, in Great Britain – melinex, in the USA – dacron, mylar, in France – terfan, estar, etc., in Belgium – gevar, in Japan – tetorone, in Germany – hostafan, in Italy – motivele. It is a polyether of terephthalic acid and ethylene glycol, which can be obtained from various derivatives of terephthalic acid and ethylene glycol [4].

The main structural unit of the macromolecule of high-temperature resistant PET is:

$$[- O - CH_2 - CH_2 - O - CO - C_6H_4 - CO -]_n .$$

Polyethylene terephthalate is produced from petroleum products using similar technology by passing through several stages of heating and cooling. – O – CO – polar group, regularly located in the chain, enhances the intermolecular interactions and forms the rigidity of the material. Compared to polyimide, polyester films have a higher crystallinity of the structure [5].

To obtain the specified strength characteristics, the amorphous film is subjected to biaxial orientation, keeping it in a stretched state at elevated temperature for a certain time, which allows achieving crystallization of the polymer and stabilization of its mechanical properties. PET products are highly durable. Ongoing research is aimed at the increase of the elastic-strength properties of polyethylene terephthalate by implementing technological processes for manufacturing products and by creating a more advanced, minimally diverse chemical structure [6]. The properties of PET are modified by introducing the derivatives of aliphatic and aromatic dicarboxylic acids, hydroxy acids, polyalkylene glycols, branched diols, substituted amines, etc. as comonomers in the process of synthesis. To ensure the best mechanical, physical, electrical properties of PET, the material is filled with various additives (glass fiber, molybdenum disulfide, fluoroplast) [7].

Radiation exposure also leads to modification of the substance. The restructuring of the supramolecular structure leads to a variety of properties of the polymer material, leading to an improvement or deterioration of its specific technical characteristics [4].

It is known that the mechanical properties of a polymer depend on its nature. Therefore, for polymers containing a benzene ring in the side chain, the degree of various changes is more significant (modulus of elasticity, creep rate, reduction of long-term strength, etc.) [4].

This work is devoted to the study of the properties of polyethylene terephthalate films of various brands under the influence of electron irradiation.

2. Experimental technique

In the work, polyethylene terephthalate films of various manufacturers were investigated: Mylar (USA) and PET (Russia).

Mechanical tests were carried out on a tensile testing machine of the RMU-0.05-1 type with a sliding speed of clamps of 36.09 ± 0.05 mm/min. Film samples were tested for uniaxial tension under normal conditions up to rupture (at T = 20 ± 2 °C, relative air humidity (45 ± 5)%). The studied samples were strips 5 x 50 and 10 x 100 mm in size, respectively, with a thickness of 100 μm.

The samples were irradiated in air in special holders on an electron accelerator of the ELU-6 type at 20 °C with an energy of 4 MeV, a current density of 0.5 $\mu A/cm^2$, a pulse duration of 5 μs at a repetition rate of 200 Hz. Absorbed doses (D) were $0 - 10^{10}$ Gy.

3. Results and discussion

Figure 1 shows the graphs of the relative elongation and stress versus electron dose for polyethylene terephthalate films of various brands. As can be seen from the figure, the nature of the curves is similar and consists of 2 stages: 1) at $D = 0 - 10^4$ Gy, both for deformation and stress, the changes are minor and slowly decrease. The relative elongation of Mylar-type films to doses of ~ 10^4 Gy and for PET to doses of ~ 10^7 Gy grows, which is associated with crosslinking processes. The second stage for a Mylar film lies in the range of $10^4 - 10^6$ Gy, where there is a sharp drop of ε and σ to 0. For Mylar films, Stage 2 is in the range of $10^7 - 10^9$ Gy.

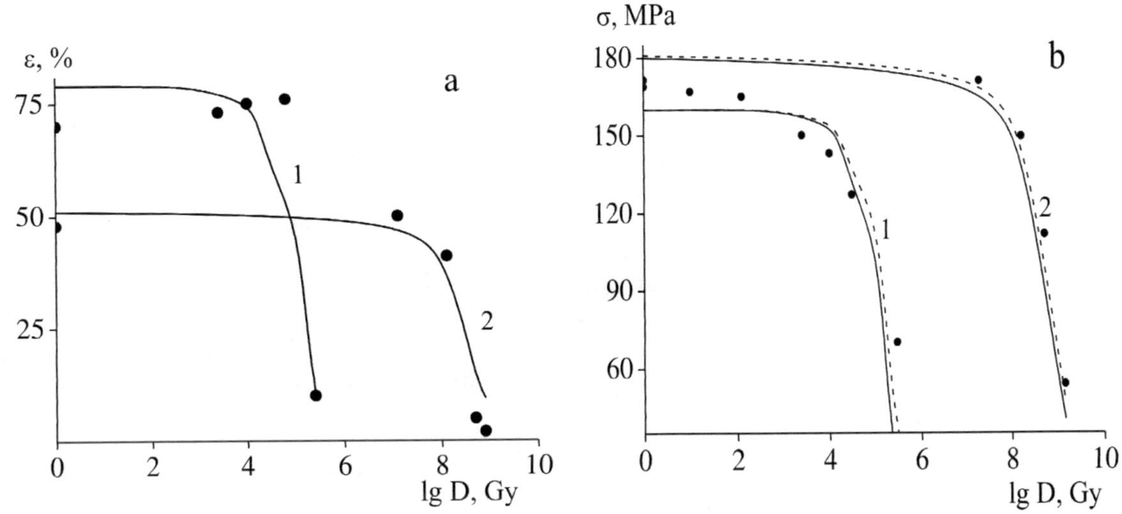

1 – Mylar (USA); 2 – PET (Russia)
points – experiment, solid line – calculation by the exponential model, dotted line – calculation by the generalized model

Figure 1. Dependences of ε (a) and σ (b) of polyethylene terephthalate films of various brands on the dose of electronic irradiation

We have considered 2 models: exponential and generalized.

The simple physical model is used for constant static load changes for the interval σ from σ_0 to σ, ε from ε_0 to ε and D from 0 to D. Then we have:

$$\varepsilon = \varepsilon_0 \, exp\left(-\frac{D}{D_0}\right), \tag{1}$$

$$\sigma = \sigma_0 \, exp\left(-\frac{D}{D_0}\right). \tag{2}$$

To obtain the generalized exponential model, we used the formula from [8]:

$$\sigma^* = \sigma_0 \ln(\varepsilon + 1). \tag{3}$$

Substituting (1) into (3) we get:

$$\sigma^* = \sigma_0 \, ln\left(\left(\varepsilon_0 \, exp\left(-\frac{D}{D_0}\right)\right) + 1\right). \tag{4}$$

The calculated dependences of ε and σ on D, obtained using formulas (1 – 4) are shown in Figure 1. As follows from the results, the theoretical curves describe the experiment satisfactorily. In this case, the generalized and exponential models give almost identical results.

Analysis of the obtained results shows that, according to the deformation properties, Mylar films have an elongation of ~ 70 %, for PET (Russia) ε is 48 %. By its strength properties, it is more resistant to electronic effects and is destroyed at doses of 10^9 Gy of PET produced in Russia, whereas Mylar's are at 10^5 Gy.

The effect of irradiation on the strength and inelastic characteristics of polymers is significant and affects the amorphous phase of the material [9 – 11]. The structure of the Mylar has a complex heterogeneous structure of amorphous regions, possessing a significant proportion of interfibrillary amorphous regions separating adjacent microfibrils, along with intrafibrillary amorphous layers [3, 4]. Basically, destruction occurs on the main chain, since high stresses occur in it.

The stress created by external loads propagates from the point of application of force along the sample at a certain rate. The weakened fragment of the film closest to the point of application of the load will be the point of rupture. The mechanical strength of the polymer is dramatically reduced by fragments of the introduced elements, which are loosely coupled with each other, which remain at very high doses.

For Russian-made PET films, this behavior of mechanical properties is associated with the features of film synthesis. Modification of properties in the process of producing polyester films is carried out by dimethyl adipate, dimethyl hexahydro-terephthalate, in consequence of which they are more durable, less creeping, and more resistant to repeated deformations [1, 2].

4. Conclusion

1. For 2 types of polyethylene terephthalate films studied (Mylar, PET), it was obtained that the character of the change of mechanical properties with the growth of the dose of electronic radiation is similar. Increasing the dose to 10^4 Gy leads to a slow change in the deformation and stress for Mylar, and a further increase to 10^6 Gy leads to a sharp decrease in them, which is associated with an increase in the destructive processes in the material. For lavsan films under irradiation to doses of ~ 10^7 Gy, a gradual decrease in ε and σ, is observed, and a further increase to 10^9 Gy leads to a sharp decrease due to the features of the synthesis technology of this material.

2. Calculations were made according to the previously proposed models. The exponential physical model on the dependence of strain and stress on the electron dose of radiation is in satisfactory agreement with experimental data. Good agreement between the calculated and experimental data was also obtained for the generalized model.

Acknowledgments

The study was supported by the Grant of the Ministry of Education and Science of the Republic of Kazakhstan (a. № 200).

References

[1] White J L, Choi D, 2006 *Polyethylene, polypropylene and other polyolefins* (Spb.: Professiya)

[2] D Brooks, J Giles, 2006 *Manufacture of packaging of PET, transl. from English* (ed. by O.Y. Sabsay, SPb: Professiya)

[3] Ardekani S M, Dehghani A, Maadeed M A, Wahit M U, Hassan A, 2014 Mechanical and thermal properties of recycled poly(ethylene terephthalate) reinforced newspaper fiber composites *Fibers Polym.* **15** 1531-1538

[4] Layus L A, Slutsker A I, Hoffman I V, Gilyarov V L, 2004 Interrelation of characteristics of reversible thermal and force deformations in solids of different structure *Physics of a solid* **46** 6 1115-1122

[5] Kudashev S V, Urmantsev U R, Seleznev G V, Rakhimov N A, Zheltobryukhov V F 2012 Surface modification of PET granules by fluoride-containing urethanes *Journal of Applied Chemistry* **85** 11 1860-1866

[6] Rouhi S, Alizadeh Y, Ansari R 2014 On the interfacial characteristics of polyethylene/single-walled carbon nanotubes using molecular dynamics simulations *Applied Surface Science* **292** 958-970

[7] Gerasimenko N N, Voronova N A, Kovtunets V A, Taipova B G, Kupchishin A I, Omarbekova Z A 2003 Study of the mechanical properties of polymer films effecting on their by static load and temperature *Proceedings of the universities. Materials of electronic technolog*y 6-12

[8] Kupchishin A I, Taipova B G, Kupchishin A A, Voronova N A, Kirdyashkin V I, Fursa T V 2016 Catastrophic models of materials destruction *IOP Conf. Series: Material Science and Engineering* **110** 012037

[9] Kupchishin A I, Niyazov M N, Voronova N A, Kirdyashkin V I, Abdukhairova A T 2017 The effect of temperature, static load and electron beam irradiation on the deformation of linear polymers *IOP Conf. Series: Materials Science and Engineering* **168** 012016

[10] Voronova N A, Kupchishin A I, Taipova B G, 2018 Nanoclusters and Electron Irradiation Effect on Mechanical Properties of Polyimide-Based Composite *Key Engineering Materials* **769** 72-77

[11] Voronova N A, Kupchishin A I, Niyazov M N, Lisitsyn V M, 2018 Uniaxial Stress and Electron Irradiation Effects on Nanochains Straightening in Film Polymer Materials *Key Engineering Materials* **769** 78-83

Experimental studies of the possible piping installation for ship systems

T M Do

Vietnam Maritime University, 484 Lach Tray Str., Haiphong City, Vietnam

E-mail: dotatmanh@vimaru.edu.vn

Abstract. Improvement in the pipelines processability of ship systems at the stage of design and technical preparation with the possibility of making pipes without local dimensioning is considered in the current research work. In the course of the experimental calculation, the objective, the tasks and the plan of the experiment were developed; the necessary initial data were prepared. According to the experimental calculation result, the theoretical predictions were confirmed for the possibilities of assembling the pipeline routes at the design stage and the conceptual basis for eliminating total deviations by the routes transition. An algorithm has been developed for eliminating deviations of pipeline routes using mutually straight sections with pipe junctions and additional allowances. According to the research results, the prerequisites for the method development to improve the technological effectiveness of pipelines at the stage of design and technological preparation were created. With regard to a greater number of pipelines, it is possible to replace the inserting pipes with fitting pipes, which will be flexible according to the design dimensions without specifying the place, helping to reduce the time required for the construction of objects saturated with pipelines.

1. Introduction

The research results in [1-3] state as follows: by turning the straight pipes and using the loose couplings, it is possible to exclude actual deviations in the working process of the piping systems. However, these results are based only on the mathematical models, as well as simulations of the geometry. The research within the real experiment is needed to accept the results. Also to gain new knowledge to exclude deviations of ship pipelines routes in the conditions of installation connections in the mutual theoretical position and the permissible nonpendicularity with respect to the pipe axis [4, 5].

The experimental studies purpose is to verify the possibility of using straight pipes to ensure the pipeline routes assembling and to confirm the developed mathematics [1-3]. If for the majority of rigidly fixed junctions the area of actual deviations turned out to be less or equal to the attainability set, calculated for a specific pipe route, then the developed theoretical positions are consistent with practice (Figure 1a). And on the contrary, if the area of actual deviations is outside of (more than) the computational space of the attainability set (Figure 1b) it means that the proposed method is not possible, it is necessary to revise the theoretical foundations in combination with other methods of eliminating deviations [6-11].

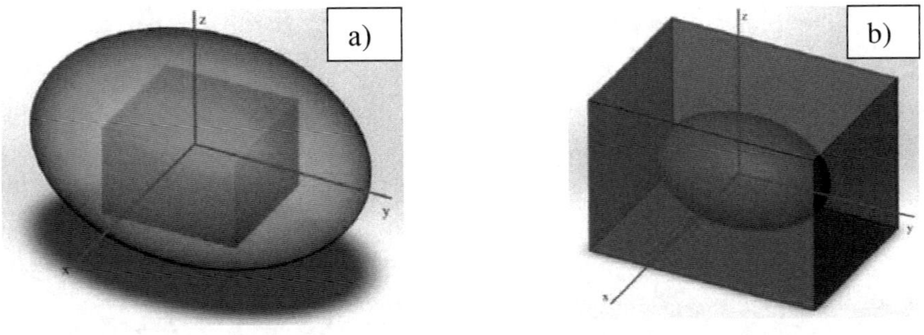

Figure 1. Comparison of deviation domain (parallelepiped) with the attainability set (ellipsoid): a) full attainability set – attainability set assimilates the deviation domain; b) attainability set assimilation of the deviation domain

2. Materials and methods

Experimental studies were carried out at the leading shipbuilding enterprises of Vietnam. The main stages of experimental studies were carried out on the real pipelines of the ship systems of the Azimuth Rotor Tug 85-32W (project No. YN-512546) (Figure 2) in the process of their production at the shipyard *LILAMA 69-2* and *Damen Song Cam Shipyard* in Vietnam.

The results of the experimental studies are the basis for the development of the methodological framework, its large-scale application in the production process of ship piping systems.

Figure 2. «Azimuth Rotor Tug 85-32W»

Main characteristics of the ship are noted in Table 1:

Table 1. Main characteristics of «Azimuth Rotor Tug 85-32W» ship.

Length by CWL, m	31.5
Width, m	13.7
Board Height, m	4.8
Draft by CWL, m	4.5
Deadweight, t	500
Speed, knot	12.5

IOP Conf. Series: Materials Science and Engineering **754** (2020) 012016 doi:10.1088/1757-899X/754/1/012016

Main engine power, kW	3 x Caterpillar 3512CTA (5400bkW)

Main stages of experimental research:

- choice of pipeline route between two rigid fixed junctions by the drawing and on board the ship;
- attainability set calculation and possible determination of pipeline route construction; function of fitting pipes with the technological allowance if necessary;
- drawing correction by measurement changes of fitting by allowance areas if necessary;
- production of the route according to the design information, including production of straight pipes with permissible displacement of flanges;
- measurements of actual deviations of assembled pipeline route;
- comparison of the actual deviations and computational attainability space;
- determination of the appropriate way of pipeline route replacement for actual deviation elimination;
- subsequent machining of the fitting pipe and the final installation.

3 Results and discussion

On the basis of the developed theoretical thesis [1] the calculation methods were worked out. As an example of calculation the pipeline route 23-2-321L00107 was considered, that is placed in the ship pipeline fuel system (Figure 3).

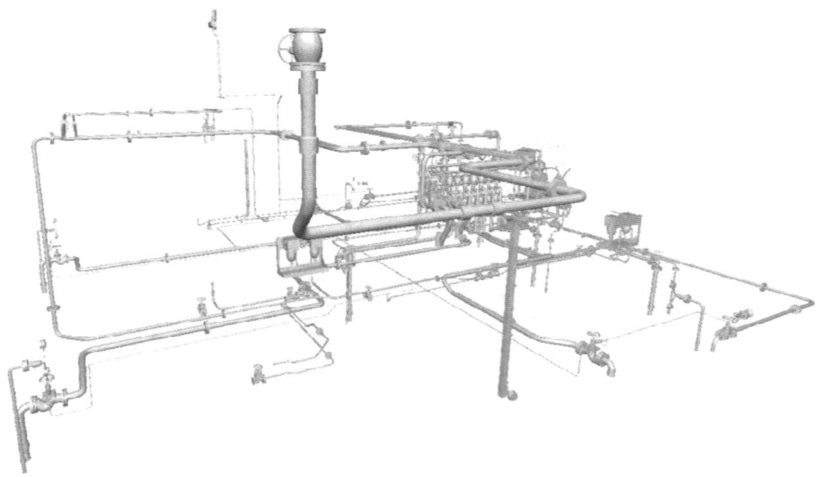

Figure 3. Ship pipeline fuel system 321

The analysis of pipeline route 23-2-321L00107 (Figure 4):

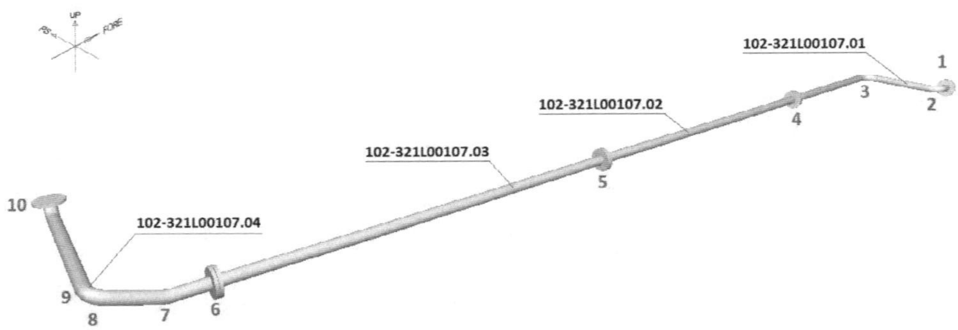

Figure 4. Calculated route 23-2-321L00107

Coordinate dimensions of the calculated route consisting of four pipes are displayed in Table 2:

Table 2. Coordination points of the calculated route.

Point number	X(FORE)	Y(PS)	Z(UP)
1	0	0	0
2	-184	0	0
3	-184	805	0
4	-1397	805	0
5	-4000	805	0
6	-6753	805	0
7	-6947	805	0
8	-7240	805	125
9	-7240	2000	125
10	-7240	2000	105

Numbers of junction points: 1, 4, 5, 6, 10. Points 1 and 10 are placed on the end of the last pipe, so it is not involved in the installation and rotation of the pipes. Finally, the numbers of junction points are determined, where the free junctions are appointed: s = {4, 5, 6}.

The further analysis of the calculated route configuration is needed. The route 23-2-321L00107 consists of two straight pipes D_y65 (102-321L00107.02 and 102-321L00107.03) and two winded pipes D_y65 (102-321L00107.01 and 102-321L00107.04). It is detected that by the setting of two straight pipes 102-321L00107.02 and 102-321L00107.03 with the junction displacement the trajectory of the final point of the pipe always moves by the circular arch.

On the basis of the mathematical description [1, 3] the calculated parameters of the circle are to be determined, that are obtained by the pipe rotation with the assumed junction displacement (Table 3). The rotation of these areas under the definite turning angle makes up the arches that are necessary for the attainability set construction.

Table 3. Pipe vector parameters of the calculated route.

№	Pipe	Vector coordinates	Distance r_i, mm	D_y, mm
1	1-4	(-1397,805,0)	1612	70
2	4-5	(-2603,0,0)	2603	70
3	5-6	(-2753,0,0)	2753	70
4	6-10	(487,1195,105)	1295	70

The developed formula in [1, 3] determines the trajectory of the last A route point displacement. This trajectory states as follows:

$$\begin{cases} X = r_1\cos\theta_1 + r_2\cos\theta_2 - X_4 + X_6 \\ Y = r_1\sin\theta_1\cos\varphi_1 + r_2\sin\theta_2\cos\varphi_2 - Y_4 + Y_6 \\ Z = r_1\sin\theta_1\sin\varphi_1 + r_2\sin\theta_2\sin\varphi_2 - Z_4 + Z_6 \end{cases} \quad (1)$$

Or:

$$\begin{cases} X = 2603\cdot\cos\theta_1 + 2603\cdot\cos\theta_2 - (-1397) + (-6603) \\ Y = 2603\cdot\sin\theta_1\cdot\cos\varphi_1 + 2603\cdot\sin\theta_2\cdot\cos\varphi_2 - 805 + 805 \\ Z = 2603\cdot\sin\theta_1\cdot\sin\varphi_1 + 2603\cdot\sin\theta_2\cdot\sin\varphi_2 - 0 + 0 \end{cases} \quad (2)$$

with θ_1, θ_2 – cross angle $[0, \alpha]$; φ_1, φ_2 – turning angle $[0, 2\pi]$.
As the result of the calculation the coordinates of attainability set are determined (mm):

$$\begin{cases} X \in (-0{,}01;\ 0) \text{ mm} \\ Y \in (0;\ 124) \text{ mm} \\ Z \in (0;\ 124) \text{ mm} \end{cases}$$

If the theoretical thesis of rigidly fixed junctions of the route are considered as a central point of coordinate system the following mark can be used: X = 0 mm; Y = ± 62 mm; Z = ± 62 mm.

The border of the reachability area is the largest value by which the piping workers can move the point 10 to eliminating the route deviation. It is necessary to provide the axes direction of the final pipe area 6-10 to be coincided with the direction to the rigidly fixed subspace (equipment, fitting…) which is perpendicular to the axes.

Two straight pipes are to be chosen (102-321L00107.02 and 102-321L00107.03) as special pipes. This data must be noted in the drawing for workers to produce the straight pipes 102-321L00107.02 and 102-321L00107.03 with the assumed junction deviation in workshop and to use free junctions in point 4 and 5 of the route. As magnitude OX in the direction of the attainability set is equal to zero, it is necessary to state the allowance size as 100 mm in this direction. The pipe 102-321L00107.04 is chosen for the route 23-2-321L00107 as the fitting pipe with the stated allowance 100 mm size by the direction OX in the area 6-7. The rest pipes can be produced as semi-production. The definite attainability set is displayed by the trajectory of the last point of the pipeline route (Figure 5).

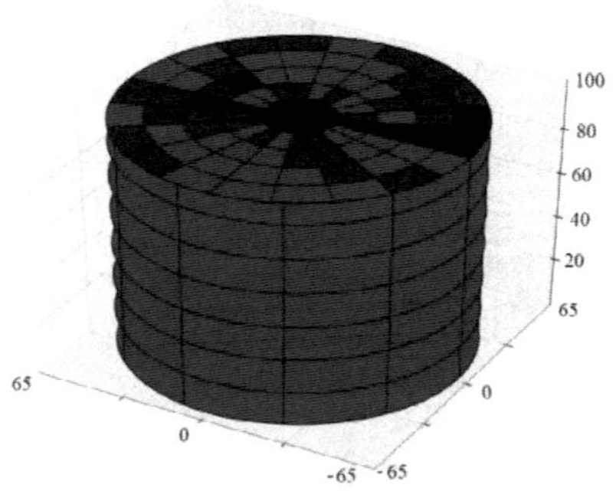

Figure 5. The attainability area of route.

Author offers the production of the straight pipes 102-321L00107.02 and 102-321L00107.03 with the assembled junction displacement on the platen: to measure the pipe length, to set the junctions, to remove junction axes by the platen with the permissible value calculated from the pipe length and fix the pipe, keeping it in the measuring position.

After that the workers shall bring all the pipes (main and fitting pipes) simultaneously to the vessel for installation. During the experiment the piping workers will measure the actual deviations in the pipeline installation in situ (Figure 6). These are precisely the quantities that must be excluded to connect the route to other routes or other equipment. In order to know whether the proposed method can be applied to exclude these deviations, the workers need to compare the point of the area of actual deviations with the calculated attainability set points.

IOP Conf. Series: Materials Science and Engineering **754** (2020) 012016 doi:10.1088/1757-899X/754/1/012016

Especially in this case the workers measure the actual deviation: ΔX = 48 mm, ΔY = 49 mm, ΔZ = 31 mm. Then they compare the points of the attainability set and see that the magnitudes of deviations in the direction of OY and OZ are less than the points of attainability set. By turning the straight pipes 102-321L00107.02 and 102-321L00107.03, produced with the admitted displacement in the free junction the piping workers can remove the final route point to the place of adjustment with the rigidly fixed junctions in the OY and OZ directions. To eliminate deviations in the OX direction, the workers cut off the excess allowance: 100 – 48 = 52 mm in the fitting pipe 102-321L00107.04. Thus, the successful installation is seen without waiting for the measurements for the production of inserting pipes as by the traditional method. As a result, the delivery time of shipbuilding orders will be reduced.

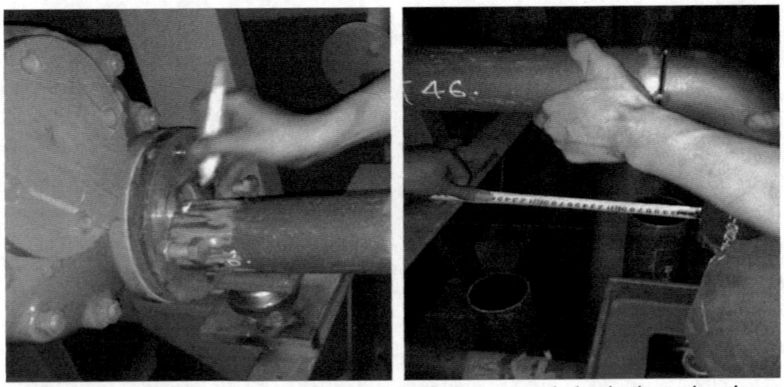

Figure 6. Taking measurements of the actual deviations in situ

Thus, by the proposed method we can completely eliminate the actual deviations. A pair of straight pipes Dy65: 102-321L00107.02 and 102-321L00107.03. Permissible displacement with the length of the specified pipes 2063: 62 mm. The pipes are made with a flange offset of 60 mm within allowance, that is, a shift of 120 mm (more than 10 mm) is carried out. In the pipeline installation process, all deviations of the welded glasses displacement are excluded.

At the stage of design and technological preparation, the analysis and determination of allowance points were carried out, pipeline drawings were corrected with indication of the points and directions of these allowances, as well as pipe sections where the allowances should be assigned. Thus, pipes can be produced according to the drawings, it is not required to take measures in situ. For the rest of the routes the appropriate calculations were carried out. According to the proposed method, 75 lines from finished pipes and fitted pipes were successfully installed according to design information without using an inserting pipe of 88 considered routes (85%). For the 13 remaining routes, it is required either to assign an inserting pipe, or use parallel pipe sections or correct their configuration.

Table 4. Fragment of experimental results.

Route name	Theoretical coordinates of the route end, *mm*		Actual deviations, *mm*	Quantity of straight pipes	Attainability set, ± *mm*	Allowance, *mm*
1-1-351L00120	X	129	11	3	78	
	Y	320	-25		67	
	Z	868	32		78	
3-1-351L00130	X	-310	-20	2	93	
	Y	888	17		93	
	Z	-4571	-32		0	100

16-2-310L00127	X	-585	-37	3	35	30
	Y	1777	19		35	30
	Z	-345	-32		78	
17-2-310L00128	X	946	26	2	26	-
	Y	2640	21		0	-
	Z	1769	25		26	-

By the theoretical calculations results of piping routes, 88 routes can be classified into 4 groups in accordance with their installation opportunities (Figure 7):

1) 19 routes (22% of the whole quantity of the analyzed pipeline routes) have got the opportunity of full deviation eliminating in three directions – the first level of possibility of piping installation (group I). All pipes of these routes can be produced according to the project data and do not require the allowance. Thus the pipeline route correction is not required in this case. But it must be noted in the drawing for the workers in the workshop to know that all pipeline routes are "finished".

2) 39 routes (44% of the whole quantity of the analyzed pipeline routes) have got the opportunity of the deviation eliminating in two directions (group II) and it is necessary to clarify the allowance in the third direction for full deviation eliminating – the second level of possibility of piping installation. All pipes of these routes can be produced according to the project data and do not require the allowance. Thus the pipeline route correction is not required in this case but it is needed to clarify the allowance on the pipe that corresponds to the necessary direction within the calculation. It must be noted in the drawing for the workers in the workshop to know that all pipeline routes are "fitted".

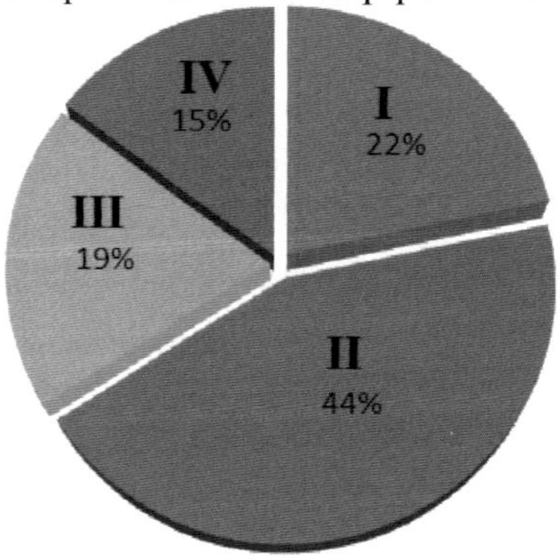

Figure 7. Classification of pipeline routes by possible installation

3) 17 routes (19% of the whole quantity of the analyzed pipeline routes) have got the opportunity of the deviation eliminating in one direction (group III); and it is necessary to clarify the allowance in the rest two directions for full deviation eliminating – the third level of possibility of piping installation.

4) 13 routes (15% of the whole quantity of the analyzed pipeline routes) do not have any opportunity of deviation eliminating – the fourth level of possibility of piping installation (group IV). In this case the pipeline routing correction is required.

The results of the experimental studies are the basis for the development of the methodological framework, its large-scale application in the production process of ship piping systems.

4 Conclusion

As a result of experimental studies of the possible assembling of straight pipes, the following was established:

- Confirmed theoretical predictions of the possible assembling of the pipeline routes at the stage of design and technological preparation;

IOP Conf. Series: Materials Science and Engineering **754** (2020) 012016 doi:10.1088/1757-899X/754/1/012016

- A procedure for eliminating deviations of pipeline routes using straight sections with pipe junctions and additional allowances has been developed;
- Prerequisites have been created for the development of appropriate methods for improving the technological effectiveness of pipelines at the design stage.

References

[1] Sakhno K N 2012 *PhD diss. Astrakhan: AGTU* 350
[2] Sakhno K N, Do T M and Vo Ch K 2016 *Mezhdunarodnyi nauchnyi zhurnal: Molodoi uchenyi. - Kazan* **6** 181-184
[3] Do T M and Sakhno K N 2018 *Vestnik Gosudarstvennogo universiteta morskogo i rechnogo flota imeni admirala S. O. Makarova* **10(2)** 356-362
[4] OST 5.95057-90 1991 *L.: NPO «Ritm»* 207
[5] RD 5R.0005-93 1994 *SPb: TSNIITS* 78
[6] Haiteng S and Wentie N 2016 *Frontiers of Mechanical Engineering* **11(3)** 214-223
[7] Asmara A 2013 *VSSD, Delft* 169
[8] Hoobasar R 2003 *Industrial Press New York* 225
[9] Lee D M, Kim S Y, Moon B Y and Kang G J 2013 *Journal of the Korean Society of Marine Engineering* **37(7)** 784-791
[10] Le L 2011 *Ha Noi: Nha xuat ban giao thong van tai* 359
[11] Fafandjel N 2008 *Urednistvo Casopisa Strojartvo* **50(3)** 141-150

IOP Conf. Series: Materials Science and Engineering **754** (2020) 012017 doi:10.1088/1757-899X/754/1/012017

Investigation of the influence of radiation on the optical biomaterials characteristics of the storage lake Sorbulak

N A Voronova[1], A I Kupchishin[1,2], M N Niyazov [1], O V Esyrev[1] and A T Abdukhairova[1]
[1] Abay Kazakh National Pedagogical University, Almaty, Kazakhstan
[2] Al-Farabi Kazakh National University, Almaty, Kazakhstan

E-mail: ankupchishin@mail.ru

Abstract. The authors carried out the study on the effects of radiation on the optical characteristics of water solutions of Lake Sorbulak during the passage of laser radiation through them. The experimental data are well consistent with the theoretical calculations of the exponential model.

1. Introduction

In recent decades, much attention has been paid to studying the conditions of accumulation of precipitation in lakes and sources of irrigation water [1 – 5]. The process of cleaning bottom sediments (BS) and biomaterials is largely determined by the rate of sedimentation and is described by the silt index, since microorganisms participate in the natural cleaning of water bodies [6, 7]. The increased silt index indicates a malfunction of the cleaning system [8 – 11].

Contaminated organic substances of storage materials lead to negative consequences for water bodies [12 – 15]. The study of the natural environment of water resources is relevant to this day. For example, the concentrations of trace elements and heavy metals in irradiated and unirradiated lake plants (basil, reed, sedge, elm, etc.) were studied in [16 – 18]. In [19] evaluated the toxicity and fertilizing ability anaerobic sludge as fertilizer after treatment with gamma radiation.

This work is devoted to the study of the effect of radiation on the transmittance of water solutions of the storage lake Sorbulak.

2. Methods of the experiment

A device was developed (Scheme 1) for the study of unirradiated and previously irradiated storage materials. It *includes* a laser source, a water solution, photodiode sensor, monitor, stand with barbell, interface, time sensor, fixing elements and data analysis block.

A water solution of the storage lake was immersed in a liquid measuring tank and placed between the light source and the recorder. The dependence of the intensity (illumination) of light (laser radiation) on time is fixed. The material was irradiated on a linear electron accelerator, where the particle energy was 4 times greater than the rest energy. Absorbed doses were 500 kGy and 3 X-rays.

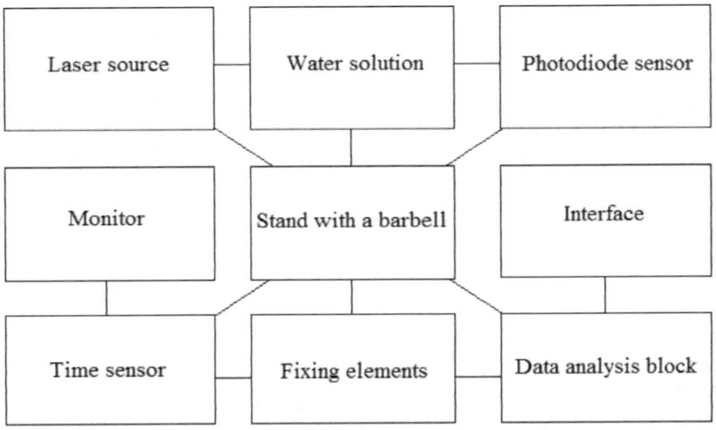

Figure 1. Picture of the experiment on the transmission of light through samples

3. Discussion of results

Figure 2 shows the dependence of the intensity of laser radiation on time.

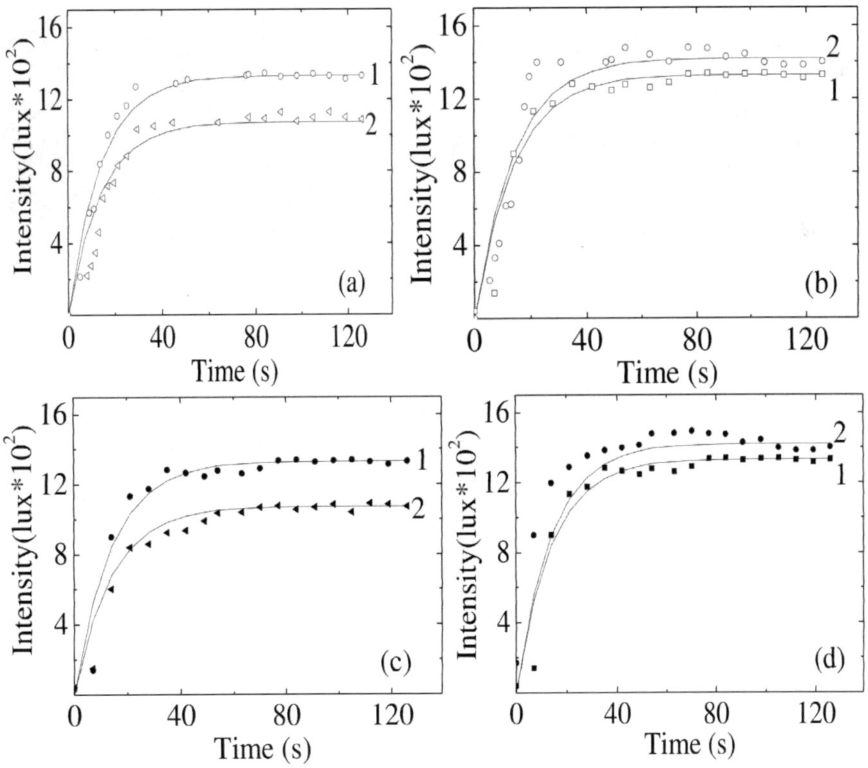

white figures – experience 2016; black figures – experiences 2018;
curves – calculation data

Figure 2. Data on the time dependences of the intensity for unirradiated (1) and irradiated by electrons (2a, c) and photons (2b, d) bottom sediments in different periods

The intensity increases over time, after which it reaches saturation. Experiments conducted on electron-irradiated deposits show that particle beams have a direct effect on the dependence I (τ). The nature of this dependence after electronic exposure is practically unchanged. The degree of natural and artificial filtration is related to the deposition rate of the accumulated material. For curves 2b, d, samples exposed to x-rays with a dose of 3P, the deposition rate decreases, which indicates the activation of microorganisms in solution.

The results obtained are satisfactorily described within the framework of the cascade-probabilistic model proposed by the authors of [20] (Fig. 2a, b, c, d):

$$I = I_0\left(1 - \exp\left(-\tau/\tau_0\right)\right),$$

here I_0 – max the illumination of the laser radiation, and τ_0 – the time during which the illumination of the material decreases by e – times.

Further, *the* study was carried out on the microstructure of bottom sediments and silt with numerous living organisms of the Lake Sorbulak (Diagrams 3, 4). As can be seen from figure 3 bottom sediments have sizes of 300 – 1000 nm, and particles of silt – 200 – 600 nm.

Figure 3. Storage materials of the Lake Sorbulak

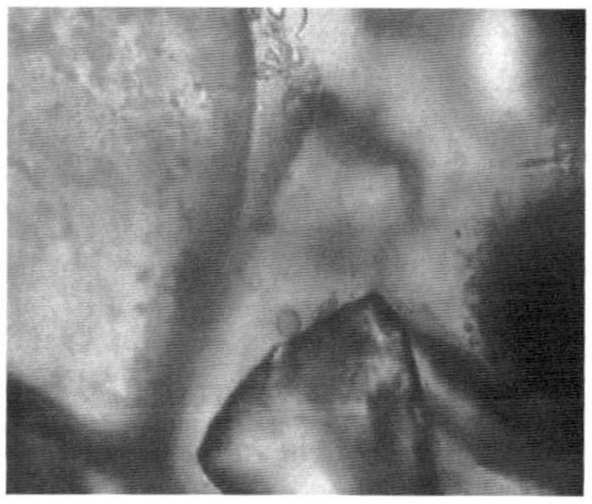

Figure 4. Silt particles of the Lake Sorbulak

4. Conclusions

1. *The* study was conducted *on* the effect of radiation on the optical characteristics of water solutions of Lake Sorbulak. The intensity of laser radiation when passing through a solution first increases sharply, and then reaches a constant.
2. At low doses of photon irradiation, a decrease in the intensity of optical radiation occurs, and at high electronic doses, an increase occurs, which is associated with an increase in the activity of microorganisms and their death, respectively.
3. The curves of the dependence of I on τ are satisfactorily described in the framework of the exponential model.
4. Microscopic studies of bottom sediments and silt with numerous living organisms of the Lake Sorbulak were carried out. As follows from the analysis of the results, bottom sediments are 300 – 1000 nm in size, and silt particles are 200 – 600 nm in size.

Acknowledgment
The study was carried out with the financial support of the grant of the Ministry of Education and Science of the Republic of Kazakhstan. IRN Grant: AP05132221.

References

[1] Babel S, del Mundo Dacera D 2006 Heavy metal removal from contaminated sludge for land appli cation: a review *Waste Manag.* **26** 988-1004

[2] Ge J, Meng X, Song Y and Terraciano A 2018 Effect of phosphate releasing in activated sludge on phosphorus removal from municipal wastewater: *Environ. Sci.* **67** 216-223

[3] Mishra D, Kim D-J, Ahn J-G and Rhee Y-H 2005 Bioleaching: a microbial process of metal recovery; a review *Met. Mater. Int.* **11** 249-256

[4] Luo D, Yuan L, Liu L, Wang Y and Fan W 2018 The Mechanism of biological phosphorus removal under Anoxic-aeration alternation condition with starch as sole carbon source and its biochemical pathway *Biochemical Engineering Journal* **132** 90-99

[5] Meulepas R J W, Gonzalez-Gil G, Teshager F M, Witharana A, Saikaly P E and Lens P N L 2015 Anaerobic bioleaching of metals from waste activated sludge *Sci. Total Environ* **514** 60-67

[6] Abu-Orf Mohammad M and Örmeci B 2005 Measuring sludge network strength using rheology and relation to dewaterability, filtration, and thickening–laboratory and full-scale experiments *J. Environ. Eng.* **131** 1139-46

[7] Herzel H, Krüger O, Hermann L and Adam C 2016 Sewage sludge ash- A promising secondary phosphorus source for fertilizer production *Sci. Total Environ* **542** 1136-43

[8] Bobade V, Baudez J C, Evans G and Eshtiaghi N 2017 Impact of gas injection on the apparent viscosity and viscoelastic property of waste activated sewage sludge *Water Res.* **114** 296-30

[9] Pittman J K, Dean A P and Osundeko O 2011 The potential of sustainable algal biofuel production using wastewater resources *Bioresour. Technol.* **102** 17-25

[10] Minh Chien Vu, Tomoaki Satomi and Hiroshi Takahashi 2020 Influence of initial water, moisture, and geopolymer content on geopolymer modified sludge *Construction and Building Materials* **235** 117420

[11] Dzyuban A N 2014 Methanogenesis and destruction of organic matter in bottom sediments of man-made water bodies *Water Resources* **41 3** 330-338.

[12] Prafulla Pokhara, Aravinth S S Ekamparam, Akhilendra B Gupta, Durgesh C Rai and Abhas Singh 2019 Activated alumina sludge as partial substitute for fine aggregates in brick making *Construction and Building Materials* **221** 244-252.

[13] Braun R, Brachtl E and Grasmug M 2003 Co digestion of protein aceous industrial waste Appl. Biochem. Biotechnol. **109** 139-153

[14] Cucina M, Ricci A, Zadra C, Pezzolla D and Gigliotti G 2019 Benefits and risks of long-term recycling of pharmaceutical sewage sludge on agricultural soil *Science of The Total Environment* **695** 133762

[15] Feinendegen L E 2016 Quantification of Adaptive Protection Following Low-dose Irradiation Health *Phys.* **110 3** 276-280

[16] Voronova N A, Kupchishin A I and Taipova B 2018 Nanoclusters and Electron Irradiation Effect on Mechanical Properties of Polyimide-Based Composite *Key Engineering Materials* **769** 72-77. doi:10.4028/www.scientific.net/KEM.769.72

[17] Behnam A L, Nosratollah N, Ebrahim M, Mohammad M and Javad H 2019 Micronutrient and heavy metal concentrations in basil plant cultivated on irradiated and non-irradiated sewage sludge- treated soil and evaluation of human health risk *Regulatory Toxicology and Pharmacology* **14** 141-145

[18] Rim D, Christelie C, Khouatmia M and Wahbi D 2018 Assessment of the toxicity and the fertilizing power from application of gamma irradiated anaerobic sludge as fertilizer: Effect on Vicia faba growth *Radiation Physics and Chemistry* **150** 163-168

[19] Kupchishin A I, Niyazov M N, Lisitsyn V M, Taipova B G, Voronova N A, Abdukhairova A T and Khodarina NN 2018 Investigation of the optical properties of unirradiated and irradiated sediments *IOP Conf. Series: Journal of Physics* **1115** 052035

IOP Conf. Series: Materials Science and Engineering **754** (2020) 012017 doi:10.1088/1757-899X/754/1/012017

[20] Yesyrev O V, Kupchishin A I, Abdukhairova A T, Nauryzbayev M K, Khodarina N N and Cherednichenko V C 2017 Study of the electron irradiation effect on the structure of treatment systems biomaterials *IOP Conf. Series: Materials Science and Engineering* **168** 012017

AUTHOR INDEX

Institute of Physics
Dirac House, Temple Back
Bristol BS1 6BE UK

ISSN: 1757-8981
ISBN 978-1-7138-2660-6

9 781713 826606

CURRAN ASSOCIATES INC.
proceedings.com